CHINESE AIR LAUNCHED WEAPONS & SURVEILLANCE, RECONNAISSANCE and TARGETING PODS

WENDELL MINNICK

EDITOR

ISBN: 9781796651638

CONTENTS

NOTE TO READER

This book is a brochure bundle.
There is no analysis. Each brochure begins with the identity of the product, the company that made it, the airshow/defense exhibition, and the number of brochures (ex. 1/4, 2/4, 3/4, etc.). This material was gathered at defense industry exhibitions and airshows in Asia, including the Zhuhai Airshow. All materials provided are for research and educational purposes.
It is raw and unvarnished.

RECOMMENDATIONS

"For the intelligence community professional, scholar, journalist and student, Wendell Minnick's books assemble years of primary documentation regarding China's military modernization. They are essential, buy them." - Rick Fisher, Senior Fellow, International Assessment and Strategy Center, Washington, DC.

"These books are a must read for anyone tracking China's military buildup, weapons technology, and arms industry. Astutely researched, Wendell Minnick's books provide never-before-revealed insights using documents that Beijing doesn't want you to see." - Ian Easton, author of book: *The Chinese Invasion Threat: Taiwan's Defense and American Strategy in Asia.*

ABOUT THE EDITOR

Wendell Minnick, BS, MA, served as the Taiwan correspondent for *Jane's Defence Weekly* from 2000-2006, Asia Bureau Chief for *Defense News* from 2006-2016, and Senior Correspondent for *Shephard News* from 2016 to 2019. Minnick now writes for *The National Interest* in Washington, DC.

Books include: *Chinese Aircraft Engines*; *Chinese Anti-Ship Missiles*; *Chinese Fighter Aircraft*; *Chinese Fixed-Wing Unmanned Aerial Vehicles*; *Chinese Helicopters*; *Chinese Radars*; *Chinese Rocket Systems*; *Chinese Rotary/VTOL Unmanned Aerial Vehicles*; *Chinese Seaplanes, Amphibious Aircraft and Aerostats/Airships*; *Chinese Space Vehicles and Programs*; and *Taiwan Space Vehicles*.

Magazine article contributions: *Afghanistan Forum*, *Apple Daily* (Chinese language), *Army Magazine*, *Asian Profile*, *Asian Thought and Society*, *Asia Times*, *BBC*, *C4ISR Journal*, *Chicago South Asia Newsletter*, *Defense News*, *Far Eastern Economic Review*, *International Peacekeeping*, *Jane's Airport Review*, *Jane's Asian Infrastructure*, *Jane's Defence Upgrades*, *Jane's Defence Weekly*, *Jane's Intelligence Review*, *Jane's Missiles and Rockets*, *Jane's Navy International*, *Jane's Sentinel Country Risk Assessments*, *Japanese Journal of Religious Studies*, *Journal of Asian History*, *Journal of Chinese Religions*, *Journal of Oriental Studies*, *Journal of Political and Military Sociology*, *Journal of Security Administration*, *Journal of the American Academy of Religion*, *Liberty Times* (Chinese language), *Military Intelligence Professional Bulletin*, *Military Review*, *New Canadian Review*, *New World Outlook*, *Pacific Affairs*, *Powerlifting USA*, *South Asia In Review*, *Taipei Times*, *Topics* (Taiwan-AMCHAM), *Towson State Journal of International Affairs*, and *The Writer.*

Covered the following defense industry exhibitions and conferences:

- Aero India – 2007, 2009, 2011
- China International Aviation and Aerospace Exhibition (Zhuhai Airshow) - 2006, 2010, 2012, 2014, 2016
- Defence Services Asia (DSA-Kuala Lumpur) - 2008, 2014, 2016, 2018
- Defence Technology Asia International Conference and Exhibit (Singapore) - 2007
- Dubai Airshow - 2007, 2009
- Global Security Asia - 2007
- International Defence Exhibition (IDEX – Abu Dhabi) – 2007
- International Maritime Defense Exhibition (IMDEX – Singapore) – 2007, 2009, 2011, 2013, 2015, 2017
- Singapore Air Show – 2006, 2008, 2010, 2012, 2014, 2016, 2018
- Seoul Air Show – 2009, 2017
- Shangri-la Dialogue (Singapore) – 2008, 2009, 2010, 2011, 2012, 2013, 2015, 2016, 2017, 2018
- Taipei Aerospace and Defense Technology Exhibition – 2001, 2003, 2005, 2007, 2009, 2011, 2013, 2015, 2017
- U.S.-Taiwan Defense Industry Conference – 2006 (Speaker)

BOOK: *Spies and Provocateurs: A Worldwide Encyclopedia of Persons Conducting Espionage and Covert Action, 1946-1991*. North Carolina: McFarland, 1992. The book was well received in the U.S. intelligence community, including positive book reviews in *Cryptolog*, *Cryptologia*, *Military Intelligence Professional Bulletin*, *Periscope* (AFIO) and *The Surveillant.* The book was also profiled in the 1995 release of the *Whole Spy Catalog: A Resource Encyclopedia for Researchers*.

ANTI-SHIP MISSILES

C705 Anti-Ship Missile Weapon System. China National Precision Machinery Import and Export Corporation (CPMIEC). 2014 Defense Services Asia (Malaysia). (1/4).

C705 Anti-Ship Missile Weapon System. China National Precision Machinery Import and Export Corporation(CPMIEC). 2014 Defense Services Asia (Malaysia). (2/4).

Main tactical & technical specifications

Typical target	1500t light frigate with a speed of not more than 35 knots
Maximum effective range	>140km
Level flight velocity	0.7~0.8 Ma
Flight altitude	(Cruise altitude) 20m
	(Level flight at terminal guidance phase) 5~7m
Guidance system	Self-control+self-guidance
	Strapdown INS with waypoints planning capability
	Basic model: 8mm active radar for terminal guidance
	Optional model: Acquisition and control command infrared imaging or TV imaging seeker for terminal guidance by replacing with the modular infrared or TV seeker and image command transmission system.
Warhead and fuze	130kg in total; one single shot can heavily damage a typical target
Propulsion system	Booster+Turbojet engine
Missile weight	Full load at stage I: 325kg
	Full load at stage II: 269kg
Overall dimension of missile	Length 4295mm (stage I)
	3555mm (stage II)
	Diameter 280mm
	Wingspan 584mm (folded)
	1600mm (unfolded)
Hit probability	90%
Launch mode	Ship-borne and vehicle-borne missile: launched from launching container
	Air-borne missile: fired from rail on the helicopter; dropped-and-fired when on the fighter plane
Launch sector	Maximum launch sector: ±90°
Operation and storage environment	Sea state: ≤5, wind speed: ≤15m/s;
	Temperature: -30℃~+60℃;
	Relative humidity: ≤ 93±3% (+40℃);

C705 Anti-Ship Missile Weapon System. China National Precision Machinery Import and Export Corporation (CPMIEC). 2014 Defense Services Asia (Malaysia). (3/4).

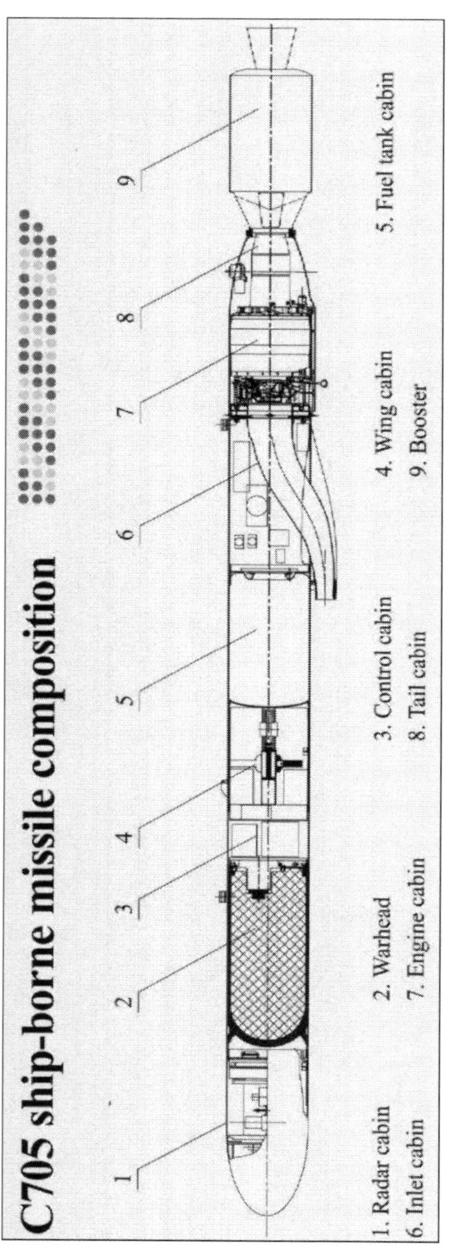

C705 Anti-Ship Missile Weapon System. China National Precision Machinery Import and Export Corporation (CPMIEC). 2014 Defense Services Asia (Malaysia). (4/4).

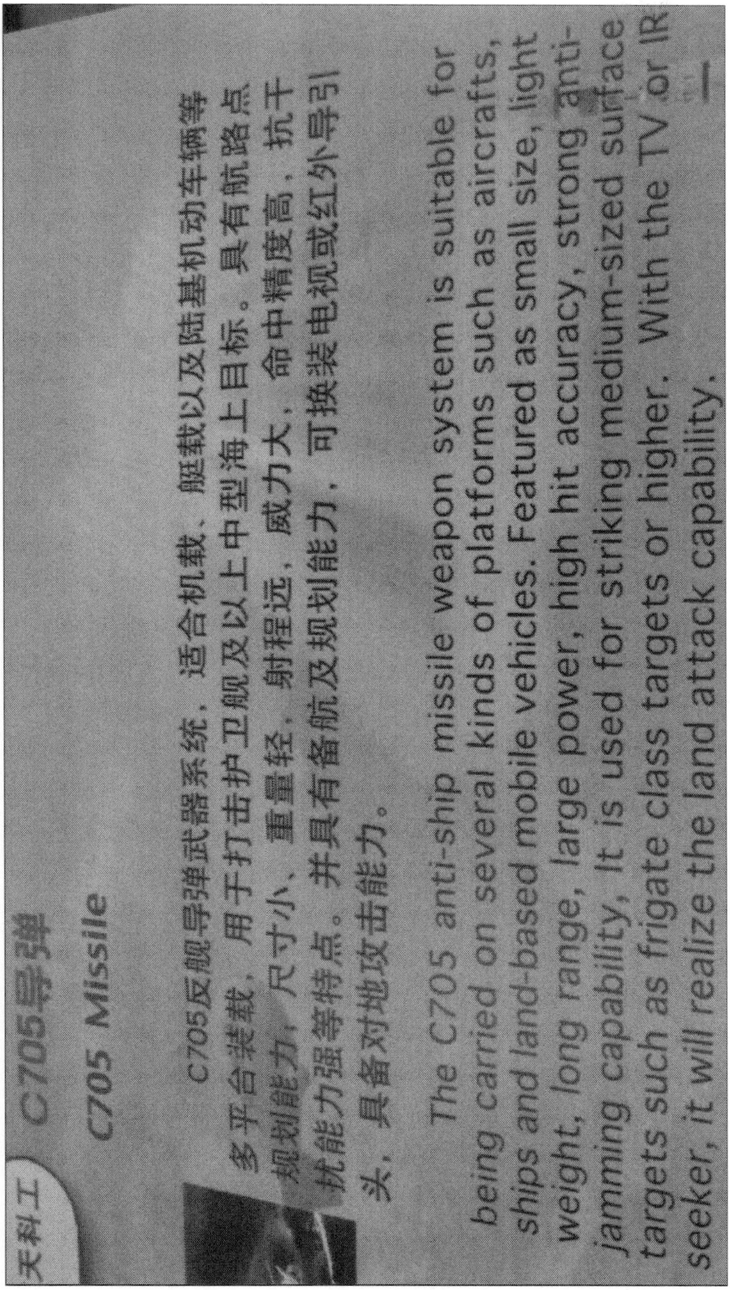

C705 Anti-Ship Missile Weapon System. Photo by Wendell Minnick. 2014 Zhuhai Airshow. (1/2).

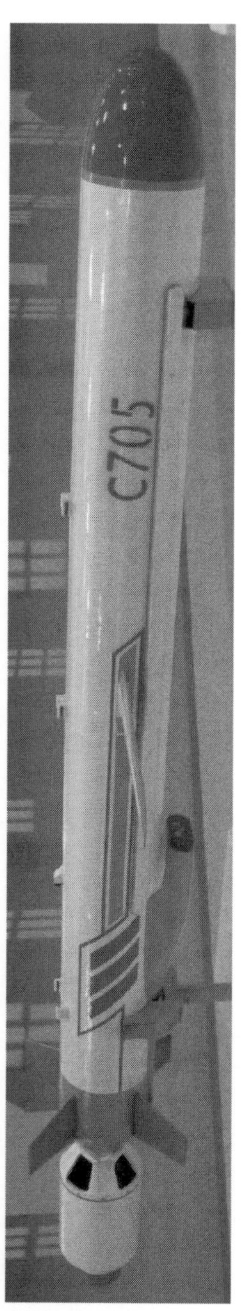

C705 Anti-Ship Missile Weapon System. Photo by Wendell Minnick. 2014 Zhuhai Airshow. (2/2).

C802 (YJ-83) Anti-Ship Missile. China National Precision Machinery Import and Export Corporation (CPMIEC). 2006 Zhuhai Airshow. (1/2).

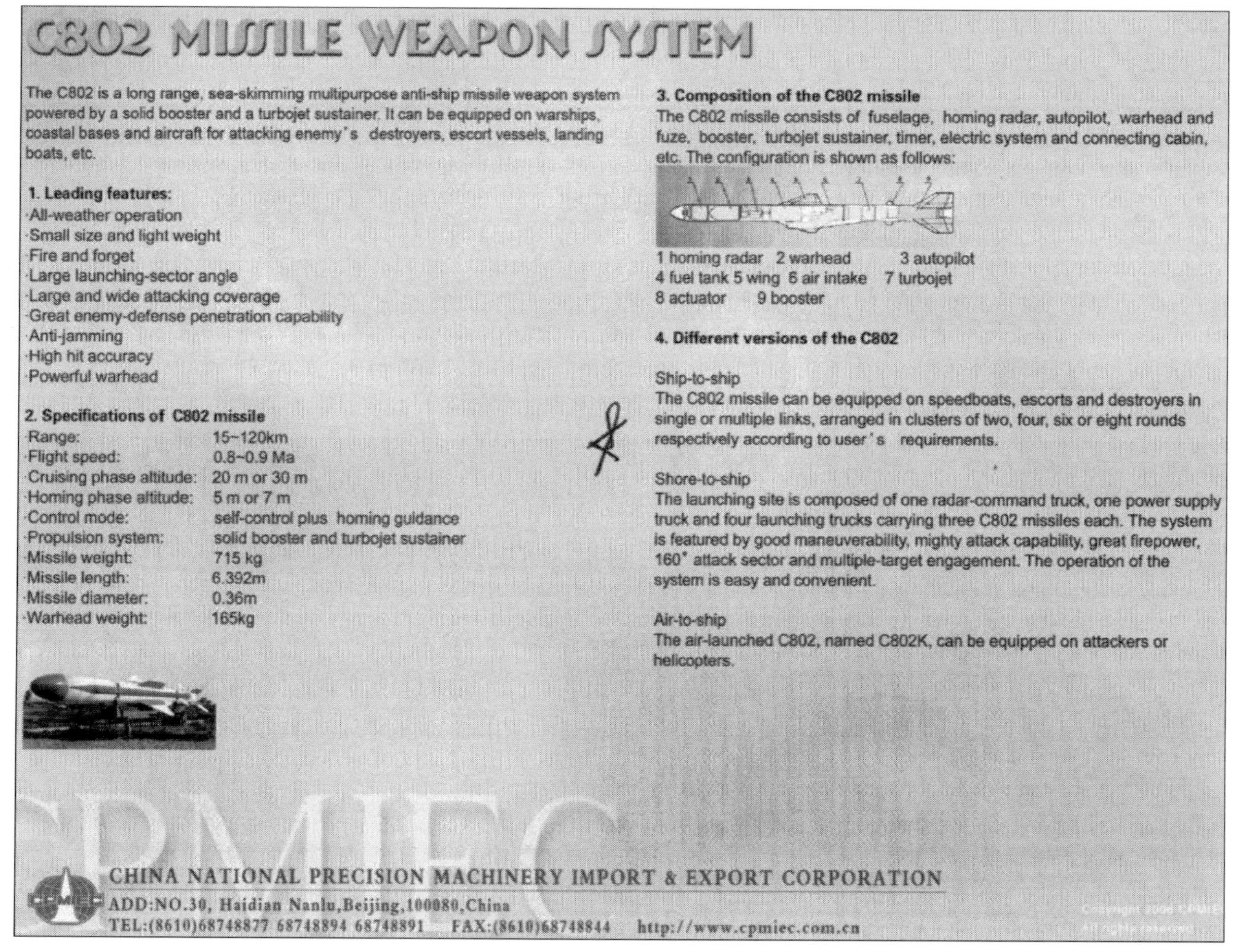

C802 (YJ-83) Anti-Ship Missile. China National Precision Machinery Import and Export Corporation (CPMIEC). 2006 Zhuhai Airshow. (2/2).

C802A Anti-Ship Missile. China National Precision Machinery Import and Export Corporation (CPMIEC). 2006 Zhuhai Airshow. (1/2).

C802A MISSILE WEAPON SYSTEM

The C802A missile is a high subsonic and sea-skimming anti-ship missile with a turbojet engine as sustainer. The main targets of C802A missile are destroyers, escort vessels, landing ships, transportation ships, etc. C802A missile can be equipped on missile boats, destroyers, escort vessels (frigates) and other ships, used as the ship-to-ship missile weapon system. It can also be equipped on the mobile launching vehicles as coast-to-ship missile weapon systems.

1. Composition

- C802A missile
- On-board (or on-vehicle) equipment
- Technical site equipment

2. Main technical specifications

- The max. range: 180 km
- The min. range: 15 km
- Flight speed: Ma=0.8 ~0.9
- Flight altitude: Cruise phase: 20m
  Terminal phase: 5m or 7m
- Launch sector angle: ± 30°
- Homing radar: Effective range: 25km
  Search angle: ± 25°
  Single-pulse active radar with frequency agility function
  Against sea-wave and with other anti-jamming measures
- Launch mode: Container launch with fixed elevation
- Launch elevation angle: 15 ± 5° (ship-to-ship missile)
  10 ± 2° (coast-to-ship missile), single launch or salvo launch

- Dimension: a. Total length: 6,383mm
  b. Diameter of missile body: 360mm
  c. Wing span: 1,220mm
- Total weight: 715kg (about)
- Control mode: Self-control + auto-guidance
- Propulsion system: One solid rocket booster+ one turbojet engine
- Fuze: Contact delay electromechanical fuze with three stage safety mechanism
- Warhead: Armor-piercing blast warhead
  Weight: 165kg (including fuze)

3. Environmental conditions:

a. Temperature: -25°C ~ +50°C
b. Relative humidity: ≤98% (+25°C)
c. Wind speed: <15m/s
d. Rain intensity: <7.5mm/h
e. Sea state: ≤6 grade

CHINA NATIONAL PRECISION MACHINERY IMPORT & EXPORT CORPORATION
Address:NO. 30, SOUTH HAIDIAN ROAD, HAIDIAN DISTRICT, BEIJING 100080, CHINA
TEX:(8610)68748877 68748891 FAX:(8610)68748844 Web site:http://www.cpmiec.com.cn

C802A Anti-Ship Missile. China National Precision Machinery Import and Export Corporation (CPMIEC). 2006 Zhuhai Airshow. (2/2).

C802A Anti-Ship Missile Weapon System. China National Precision Machinery Import and Export Corporation (CPMIEC). 2012 Defence Services Asia (Malaysia). (1/4).

C802A Anti-Ship Missile Weapon System. China National Precision Machinery Import and Export Corporation (CPMIEC). 2012 Defence Services Asia (Malaysia). (2/4).

Multiple launching platforms

Air-borne platform

Ship-borne platform

Over-the-horizon target indication

Vehicle-borne platform

Launching site deployment

C802A Anti-Ship Missile Weapon System. China National Precision Machinery Import and Export Corporation (CPMIEC). 2012 Defence Services Asia (Malaysia). (3/4).

Main tactical & technical specifications

Typical targets	5000T Class destroyer
	RCS $\geqslant 3000m^2$
Effective range	15-180km
Route planning	Max. 4 attacking paths
	Max. 3 way points per path
Flight speed	0.8-0.9Ma
Flight altitude	20m in cruising phase
	5m or 7m in terminal phase
Guidance mode	Strapdown INS+frequency agilityradar seeker for terminal guidance
Launch mode	Launching from a container 15° fixed elevation. Single launch or salvo launch
Response time	9 min in cold state
	30s in hot state
Missile dimensions	Length 6383mm
	Diameter 360mm
	Wingspan 1220mm
Weight	800kg
Fuze	electromechanical contact delay fuze
Warhead	semi-armor piercing blast warhead,190kg
Propulsion	booster + turbojet
Single-shot killing probability	90%

C802A Anti-Ship Missile Weapon System. China National Precision Machinery Import and Export Corporation (CPMIEC). 2012 Defence Services Asia (Malaysia). (4/4).

CHINA NATIONAL PRECISION MACHINERY IMPORT & EXPORT CORPORATION

C802A

MEDIUM- RANGE MULTI-PURPOSE ANTI-SHIP MISSILE WEAPON SYSTEM

As a new generation of anti-ship missile updated from C802 with strapdown INS, frequency agility radar and digital control. It features small size, light weight, long range, strong defense penetrating capability, high hitting accuracy, powerful warhead and easy operation & maintenance.

C802A Medium-Range Multi-Purpose Anti-Ship Missile. China National Precision Machinery Import and Export Corporation (CPMIEC). 2016 Defence Services Asia (Malaysia). (1/2).

C802A

MEDIUM- RANGE MULTI-PURPOSE ANTI-SHIP MISSILE WEAPON SYSTEM

1. Leading features

Multiple launching platforms
Sea-skimming flight
Multiple anti-jamming capabilities
Optimal destruction effect
Fire and forget
High performance-cost ratio
Over-the-horizon attack
Operation mode selectable for different targets

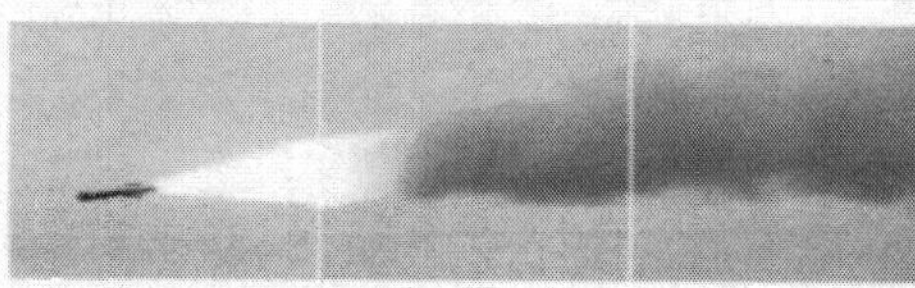

Sea skimming flight

2. Main tactical and technical specifications

Typical targets:	5,000T Class destroyer, RCS≥3,000m²
Max. effective range:	180km
Min. range:	15km
Flight speed:	0.8~0.9Ma
Flight altitude:	20m in cruising phase 5m or 7m in terminal phase
Guidance Mode:	Strapdown INS+frequency agility radar seeker for terminal guidance
Route planning:	Max. 4 attacking paths 4 way points per path
Single-shot kill probability:	90%
Warhead:	semi-armor piercing blast warhead, 190kg
Fuze:	electromechanical contact delay fuze
Propulsion system:	solid rocket booster+turbojet engine
Missile Weight:	800kg
Missile dimensions	
Length:	6.383m
Diameter:	0.36m
Wingspan:	1.22m
Launch Mode:	Launch from container at 15° fixed elevation. single launch or salvo launch

3. Missile Composition

C802A missile consists of missile body, radar seeker, strap-down INS, integrated control unit, rudder system, warhead, fuze, electronic system, turbojet engine, oiling system, boosters, etc.

4. Multiple launching platforms

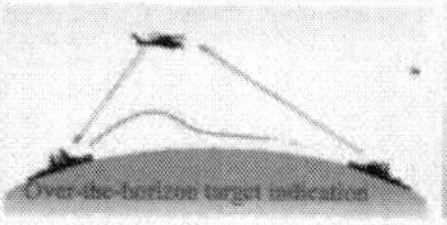

CHINA NATIONAL PRECISION MACHINERY IMPORT & EXPORT CORPORATION

ADD: No.30, Haidian Nanlu, Beijing 100080, China
TEL: 010-68748894 68748748 68748837
FAX: 010-68748844 68748880

http: //www.cpmiec.com.cn

C802A Medium-Range Multi-Purpose Anti-Ship Missile. China National Precision Machinery Import and Export Corporation (CPMIEC). 2016 Defence Services Asia (Malaysia). (2/2).

CM-302 Supersonic Anti-Ship Missile Weapon System. Photo by Wendell Minnick. 2016 Zhuhai Airshow. (1/3).

CM-302 Supersonic Anti-Ship Missile Weapon System. Photo by Wendell Minnick. 2016 Zhuhai Airshow. (2/3).

CM-302 Supersonic Anti-Ship Missile Weapon System. Photo by Wendell Minnick. 2016 Zhuhai Airshow. (3/3).

CHINA NATIONAL PRECISION MACHINERY IMPORT & EXPORT CORPORATION

DSA 2014

CM-802AKG

AIR-TO-GROUND MISSILE WEAPON SYSTEM

As a new generation air to ground missile, the CM-802AKG uses strapdown INS / satellite integrated navigation system, infrared / TV guidance seeker and digital control. The missile features small size, light weight, high hitting accuracy, good adaptability, easy & flexible combat operation and simple maintenance.

CM-802AKG Air-to-Ground Missile Weapon System. China National Precision Machinery Import and Export Corp. (CPMIEC). 2014 Defence Services Asia. (Malaysia) (1/2).

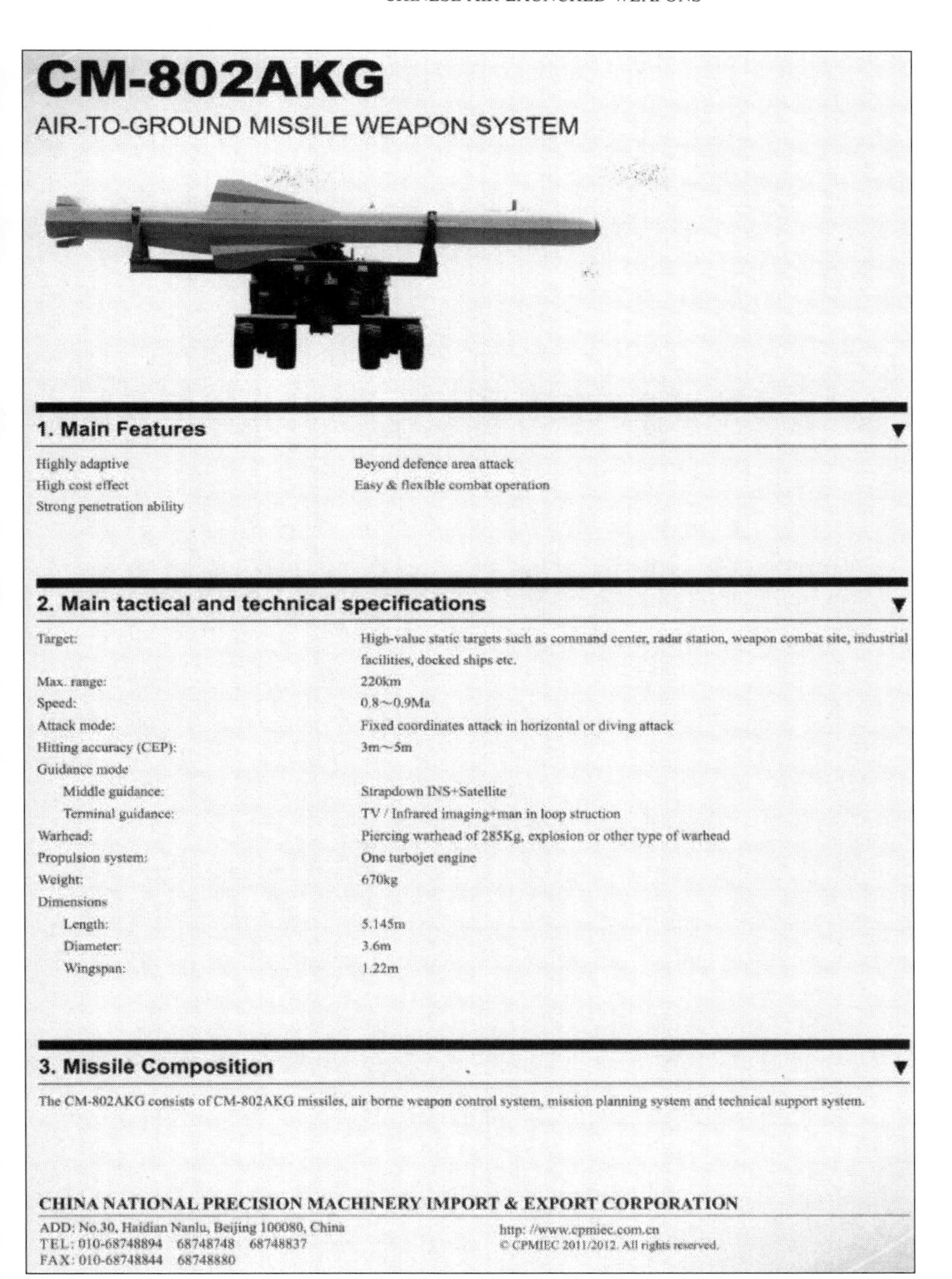

CM-802AKG

AIR-TO-GROUND MISSILE WEAPON SYSTEM

1. Main Features

Highly adaptive
High cost effect
Strong penetration ability
Beyond defence area attack
Easy & flexible combat operation

2. Main tactical and technical specifications

Target:	High-value static targets such as command center, radar station, weapon combat site, industrial facilities, docked ships etc.
Max. range:	220km
Speed:	0.8～0.9Ma
Attack mode:	Fixed coordinates attack in horizontal or diving attack
Hitting accuracy (CEP):	3m～5m
Guidance mode	
Middle guidance:	Strapdown INS+Satellite
Terminal guidance:	TV / Infrared imaging+man in loop struction
Warhead:	Piercing warhead of 285Kg, explosion or other type of warhead
Propulsion system:	One turbojet engine
Weight:	670kg
Dimensions	
Length:	5.145m
Diameter:	3.6m
Wingspan:	1.22m

3. Missile Composition

The CM-802AKG consists of CM-802AKG missiles, air borne weapon control system, mission planning system and technical support system.

CHINA NATIONAL PRECISION MACHINERY IMPORT & EXPORT CORPORATION

ADD: No.30, Haidian Nanlu, Beijing 100080, China
TEL: 010-68748894 68748748 68748837
FAX: 010-68748844 68748880

http: //www.cpmiec.com.cn

CM-802AKG Air-to-Ground Missile Weapon System. China National Precision Machinery Import and Export Corp. (CPMIEC). 2014 Defence Services Asia. (Malaysia) (2/2).

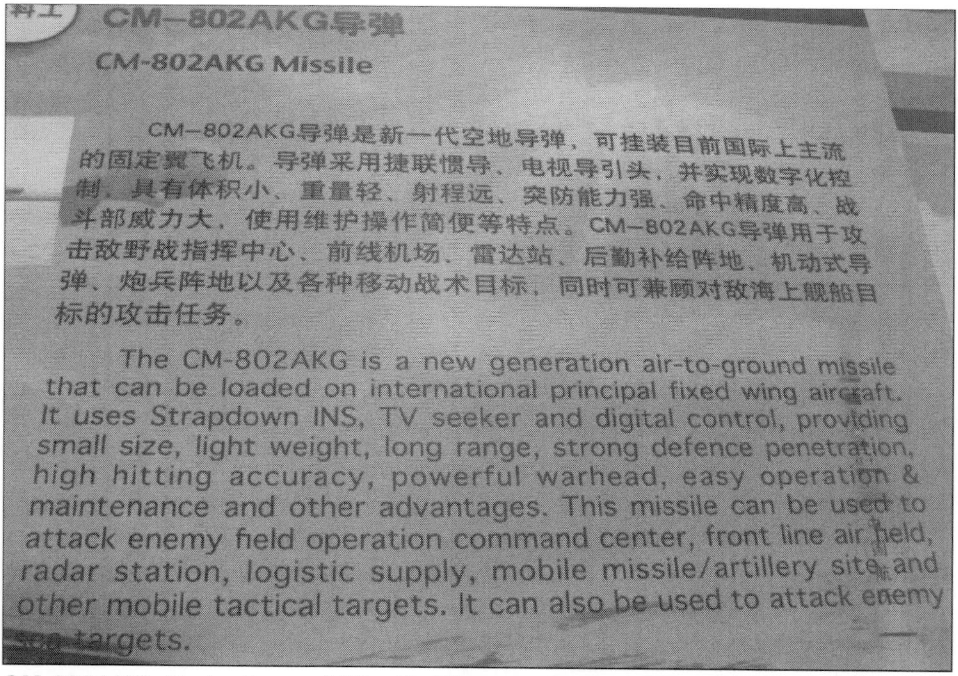

CM—802AKG导弹

CM-802AKG Missile

CM—802AKG导弹是新一代空地导弹，可挂装目前国际上主流的固定翼飞机。导弹采用捷联惯导、电视导引头，并实现数字化控制，具有体积小、重量轻、射程远、突防能力强、命中精度高、战斗部威力大，使用维护操作简便等特点。CM—802AKG导弹用于攻击敌野战指挥中心、前线机场、雷达站、后勤补给阵地、机动式导弹、炮兵阵地以及各种移动战术目标，同时可兼顾对敌海上舰船目标的攻击任务。

The CM-802AKG is a new generation air-to-ground missile that can be loaded on international principal fixed wing aircraft. It uses Strapdown INS, TV seeker and digital control, providing small size, light weight, long range, strong defence penetration, high hitting accuracy, powerful warhead, easy operation & maintenance and other advantages. This missile can be used to attack enemy field operation command center, front line air field, radar station, logistic supply, mobile missile/artillery site and other mobile tactical targets. It can also be used to attack enemy sea targets.

CM-802AKG Air-to-Ground Missile Weapon System. Photo by Wendell Minnick. 2014 Zhuhai Airshow. (1/2).

CM-802AKG Air-to-Ground Missile Weapon System. Photo by Wendell Minnick. 2014 Zhuhai Airshow. (2/2).

Air-to-ground Weapon System

TL-1 Air-to-ground Missile

As a light airborne air-to-ground guided missile system, TL-1 can be launched by UAV and helicopter, and perform accurate attack on solid fortifications, including small-sized command post, moving target on ground (such as armored cars), small-sized target on water (such as PGM missile gunboat and patrol boat), etc. The missile is characterized by flexible operation, "fire-and-forget" and "man-in-loop".

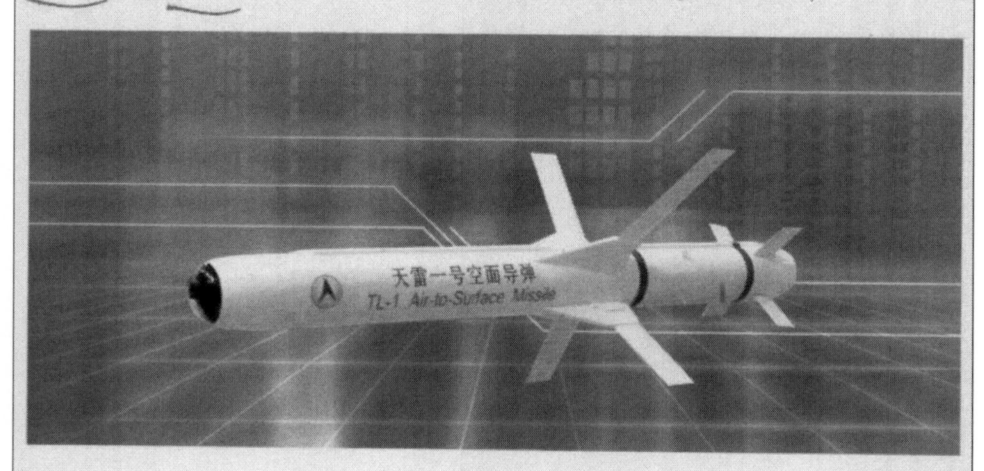

Main Features

- Compatible with UAVs and Helicopter
- Fire-and-Forget & Man-in-loop
- Dynamic trajectory planning
- High Accuracy
- Range expandable
- Modular Warhead

Technical Specifications

Range	3-20 km
Diameter of Missile	Φ180 mm
Length of Missile	1,954 mm
Maximum Speed	Mach 0.8
Guidance Mode	INS / GNSS + IR / Imaging terminal homing guidance + data link
Accuracy	CEP≤ 1 m
Weight of Mass	85 kg
Weight of Warhead	15 kg
Type of Warhead	Penetration / Fragment warhead

TL-1 Air-to-Ground Missile. ALIT/China Aerospace Long-March International. 2017 Langkawi International Maritime and Aerospace Exhibition (LIMA/Malaysia) (1/1).

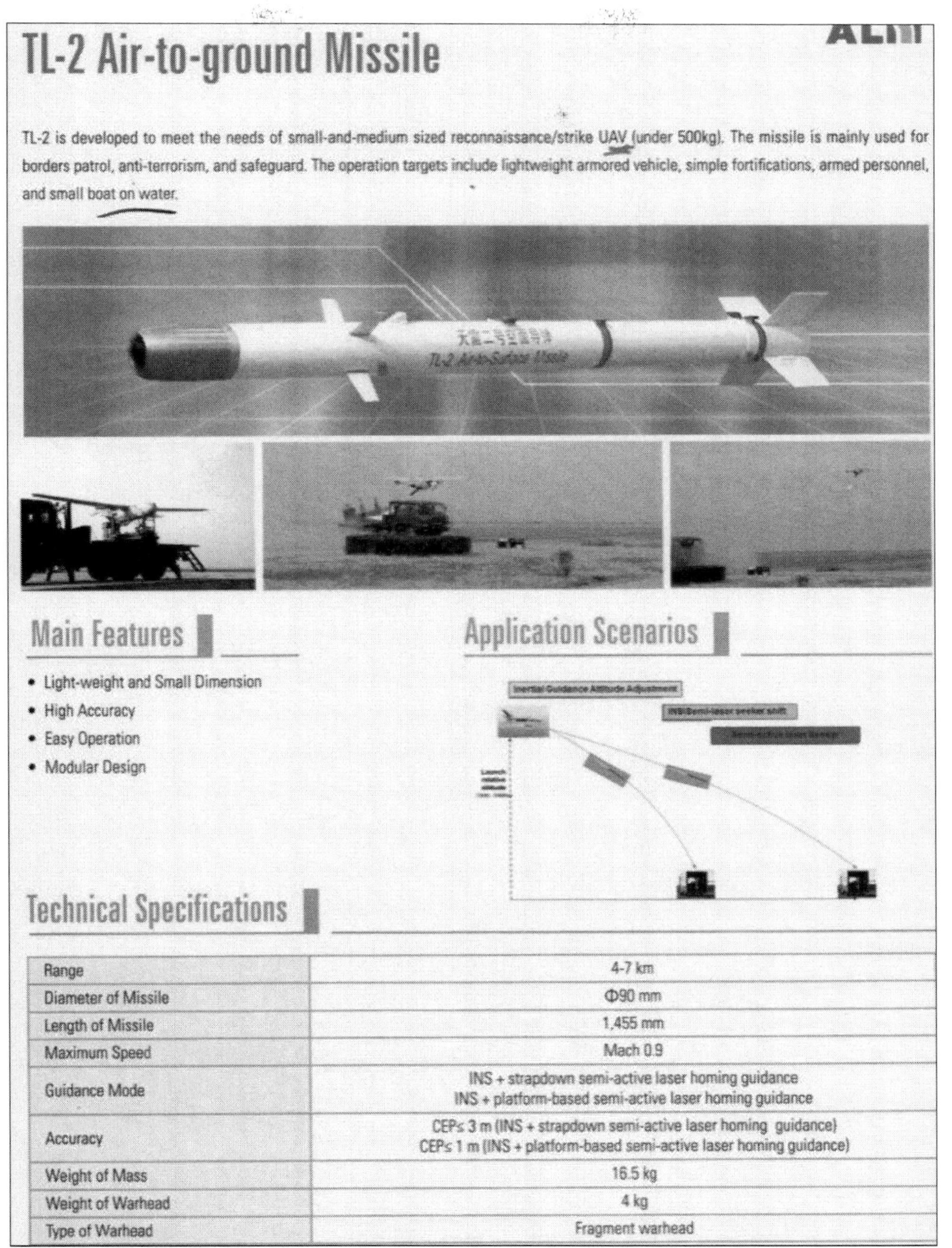

TL-2 Air-to-ground Missile

TL-2 is developed to meet the needs of small-and-medium sized reconnaissance/strike UAV (under 500kg). The missile is mainly used for borders patrol, anti-terrorism, and safeguard. The operation targets include lightweight armored vehicle, simple fortifications, armed personnel, and small boat on water.

Main Features

- Light-weight and Small Dimension
- High Accuracy
- Easy Operation
- Modular Design

Application Scenarios

Technical Specifications

Range	4-7 km
Diameter of Missile	Φ90 mm
Length of Missile	1,455 mm
Maximum Speed	Mach 0.9
Guidance Mode	INS + strapdown semi-active laser homing guidance INS + platform-based semi-active laser homing guidance
Accuracy	CEP≤ 3 m (INS + strapdown semi-active laser homing guidance) CEP≤ 1 m (INS + platform-based semi-active laser homing guidance)
Weight of Mass	16.5 kg
Weight of Warhead	4 kg
Type of Warhead	Fragment warhead

TL-2 Air-to-Ground Missile. ALIT/China Aerospace Long-March International. 2017 Langkawi International Maritime and Aerospace Exhibition (LIMA/MALAYSIA). (1/1).

TL-7 Anti-Ship Missile. China National Aero-Technology Import and Export Corporation (CATIC). 2016 Singapore Airshow. (1/2).

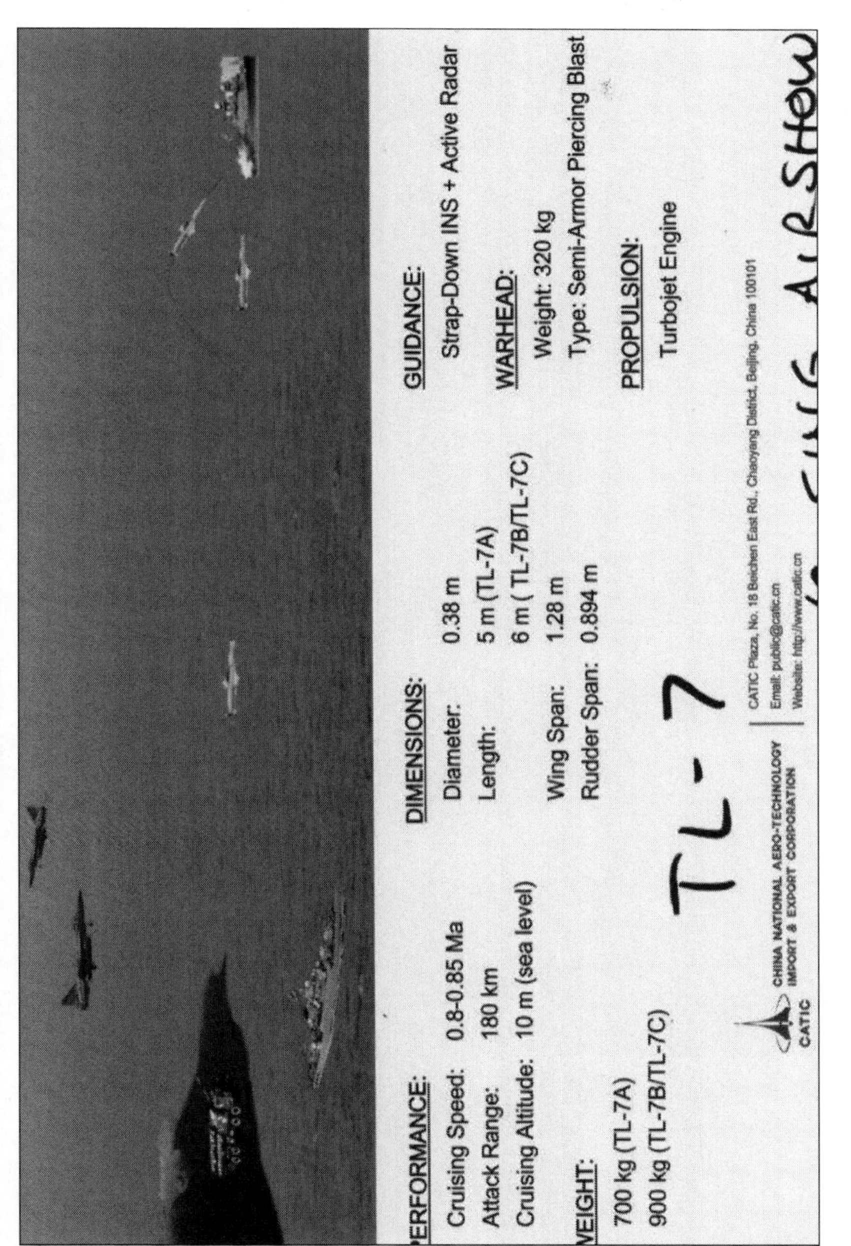

TL-7 Anti-Ship Missile. China National Aero-Technology Import and Export Corporation. 2016 Singapore Airshow. (2/2).

TL-7/17. Minnick Photo. 2016 Zhuhai Airshow. (1/2).

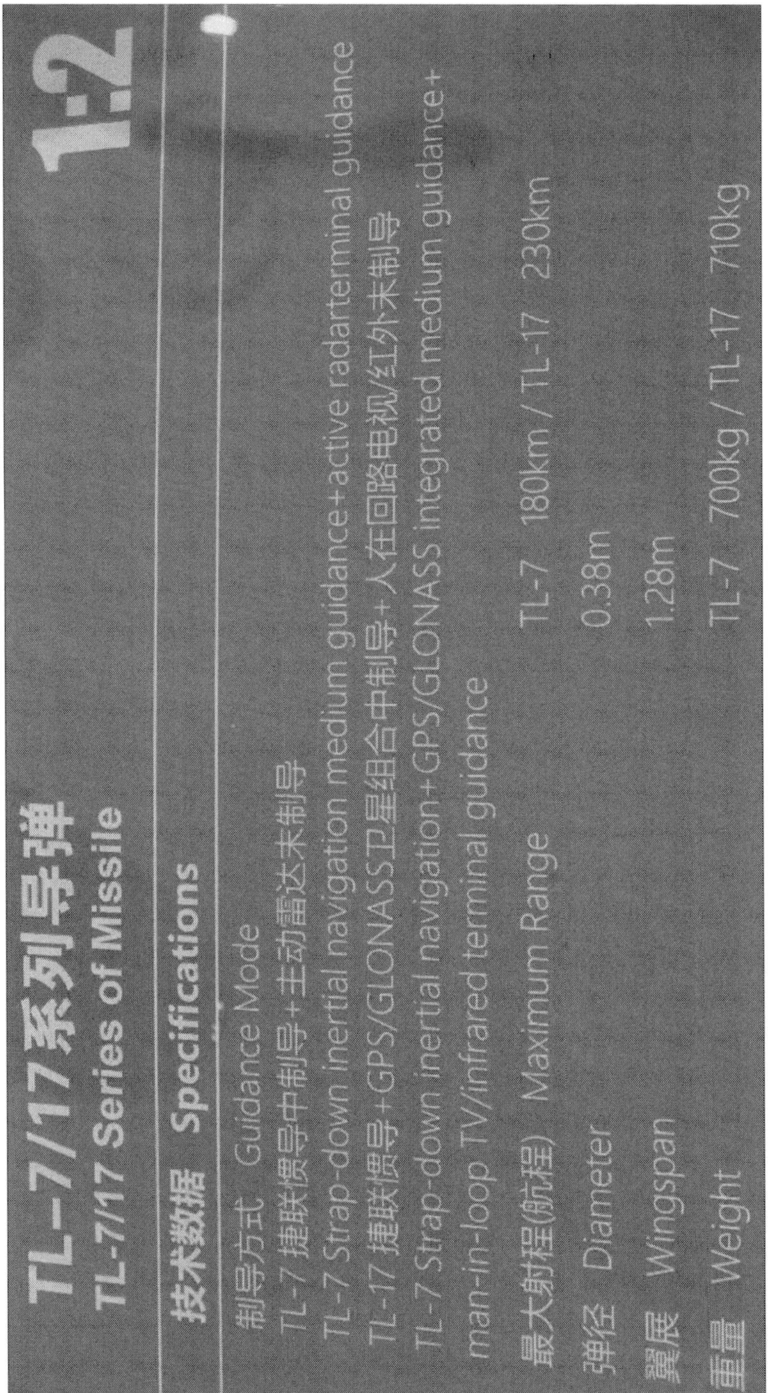

TL-7/17系列导弹
TL-7/17 Series of Missile

1:2

技术数据 Specifications

制导方式 Guidance Mode
TL-7 捷联惯导中制导+主动雷达末制导
TL-7 Strap-down inertial navigation medium guidance+active radarterminal guidance
TL-17 捷联惯导+GPS/GLONASS卫星组合中制导+人在回路电视/红外末制导
TL-7 Strap-down inertial navigation+GPS/GLONASS integrated medium guidance+
man-in-loop TV/infrared terminal guidance
最大射程(航程) Maximum Range TL-7 180km / TL-17 230km
弹径 Diameter 0.38m
翼展 Wingspan 1.28m
重量 Weight TL-7 700kg / TL-17 710kg

TL-7/17. Minnick Photo. 2016 Zhuhai Airshow. (2/2).

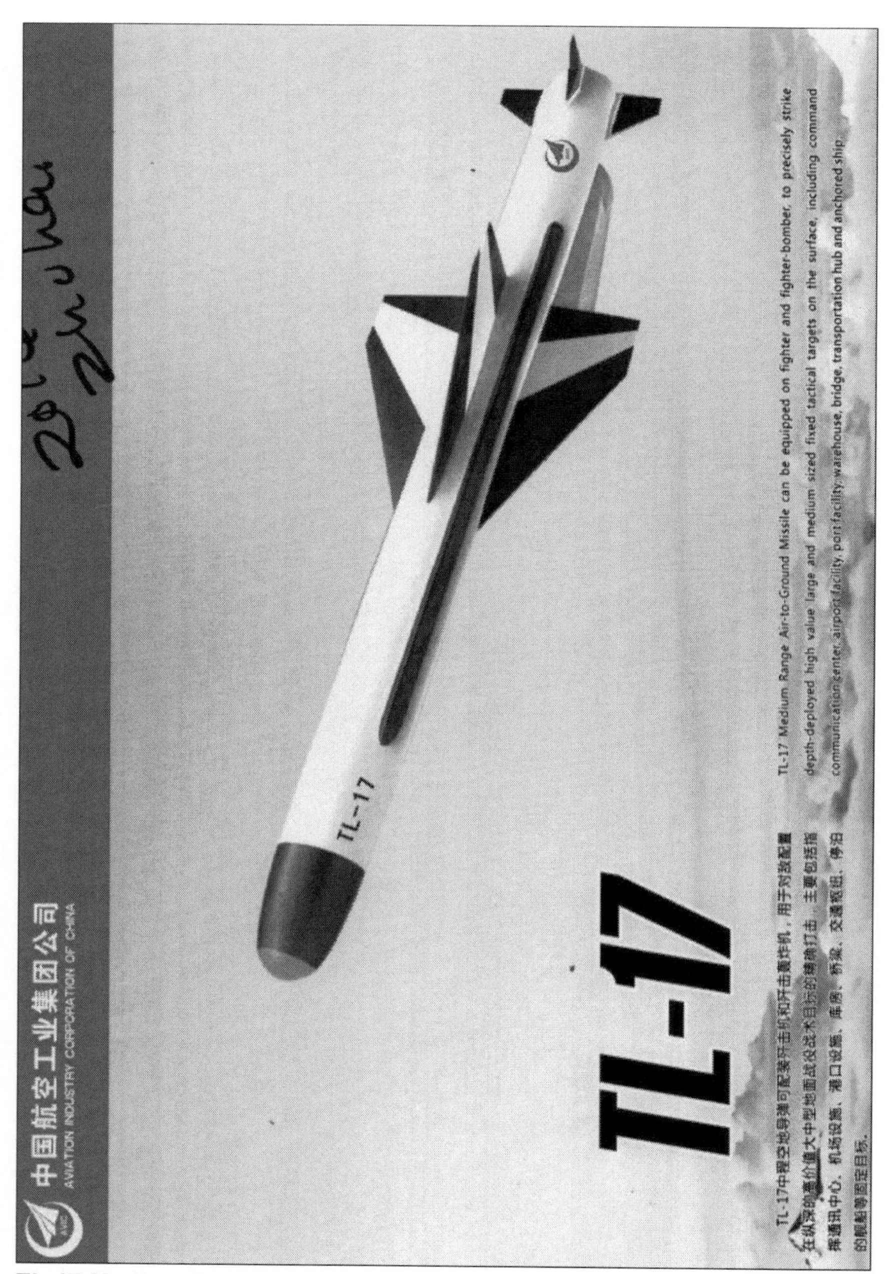

TL-17 Medium Range Air-to-Ground Missile. Aviation Industry Corporation of China (AVIC). 2016 Zhuhai Airshow. (1/5).

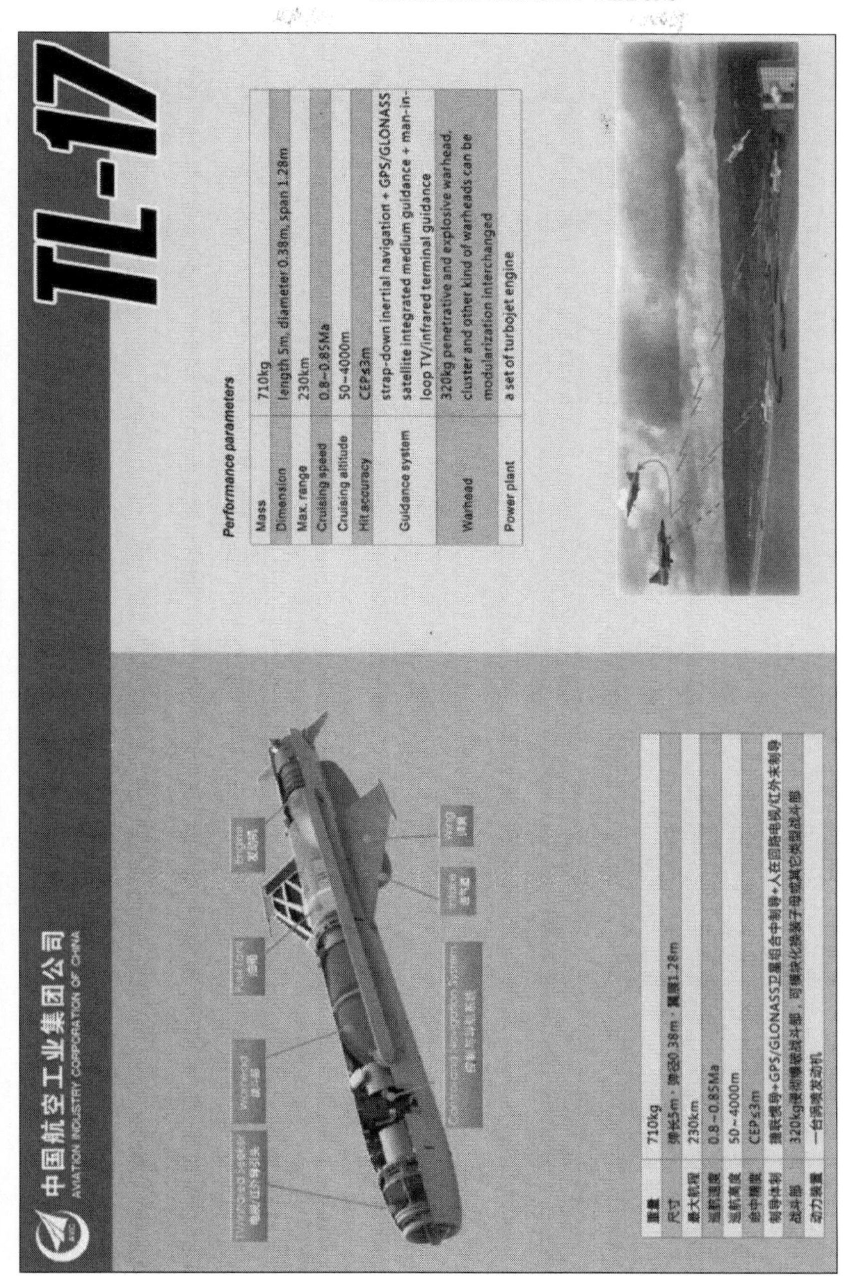

TL-17 Medium Range Air-to-Ground Missile. Aviation Industry Corporation of China (AVIC). 2016 Zhuhai Airshow. (2/5).

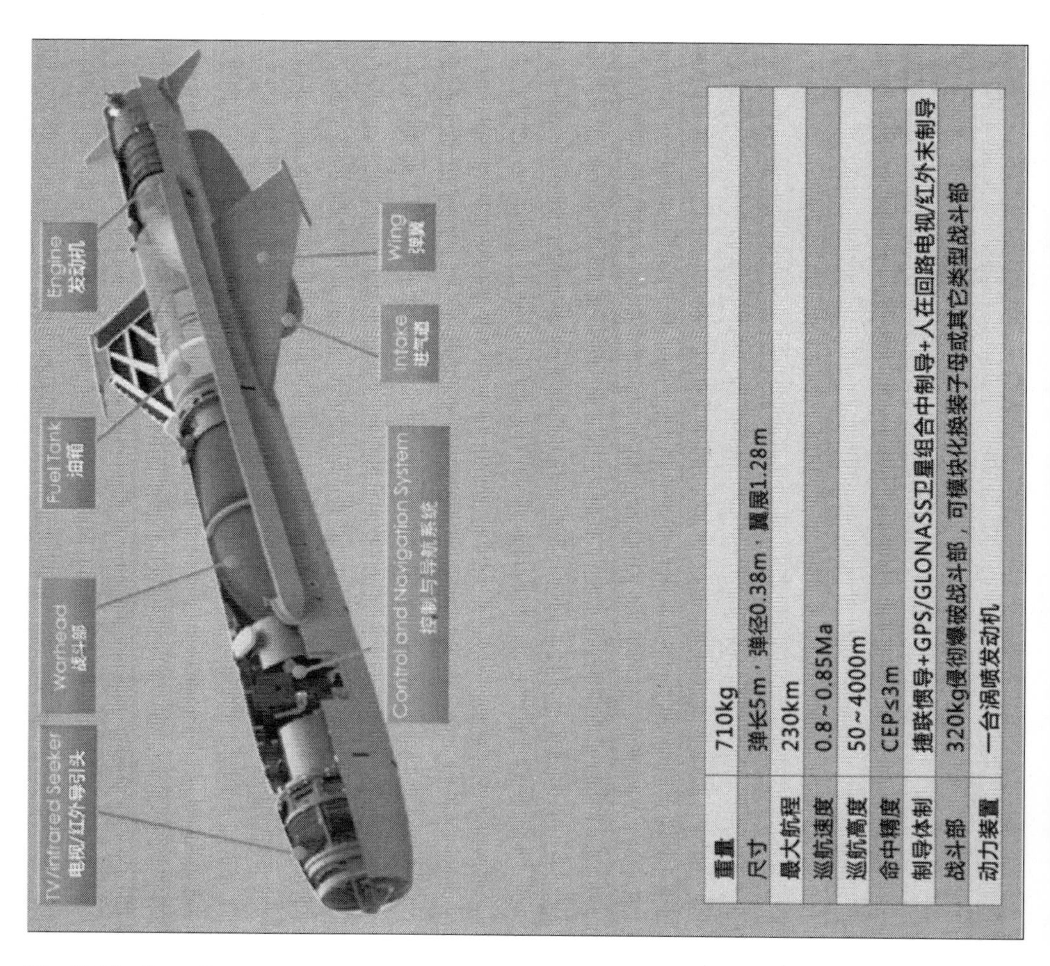

TL-17 Medium Range Air-to-Ground Missile. Aviation Industry Corporation of China (AVIC). 2016 Zhuhai Airshow. (3/5).

Performance parameters	
Mass	710kg
Dimension	length 5m, diameter 0.38m, span 1.28m
Max. range	230km
Cruising speed	0.8~0.85Ma
Cruising altitude	50~4000m
Hit accuracy	CEP≤3m
Guidance system	strap-down inertial navigation + GPS/GLONASS satellite integrated medium guidance + man-in-loop TV/infrared terminal guidance
Warhead	320kg penetrative and explosive warhead, cluster and other kind of warheads can be modularization interchanged
Power plant	a set of turbojet engine

TL-17 Medium Range Air-to-Ground Missile. Aviation Industry Corporation of China (AVIC). 2016 Zhuhai Airshow. (4/5).

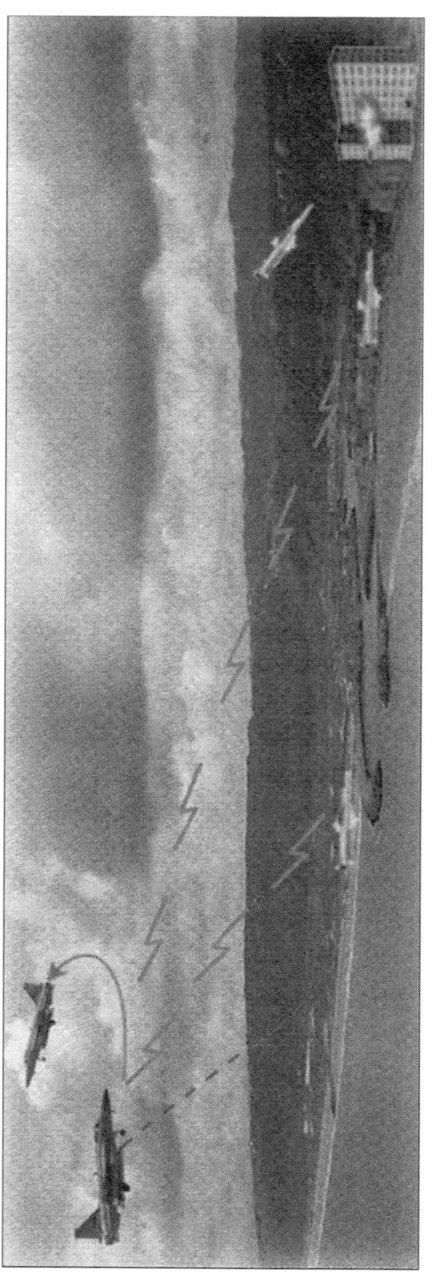

TL-17 Medium Range Air-to-Ground Missile. Aviation Industry Corporation of China (AVIC). 2016 Zhuhai Airshow. (5/5).

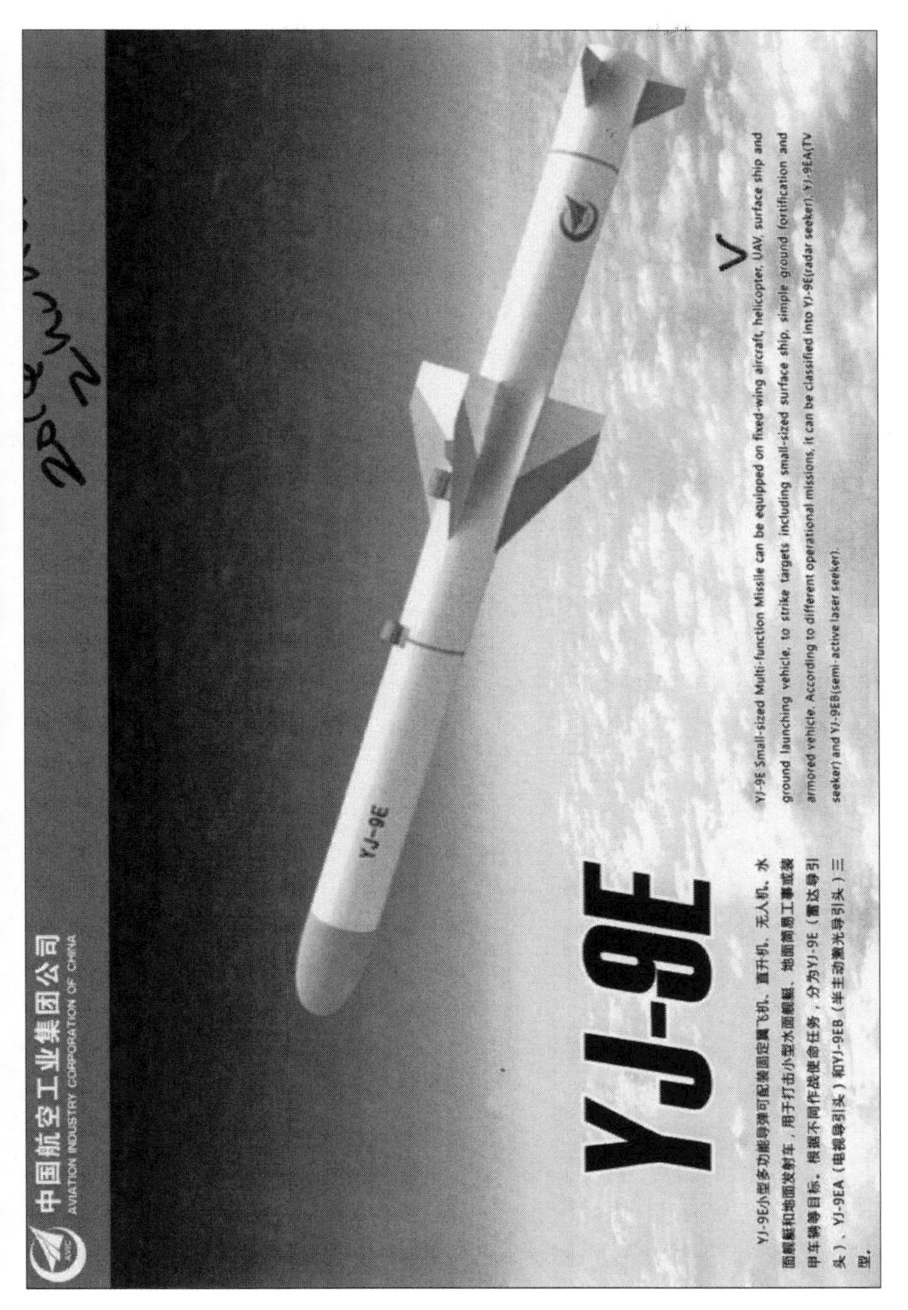

YJ-9E Small-Sized Multi-Function Missile. Aviation Industry Corporation of China (AVIC). 2016 Zhuhai Airshow. (1/5).

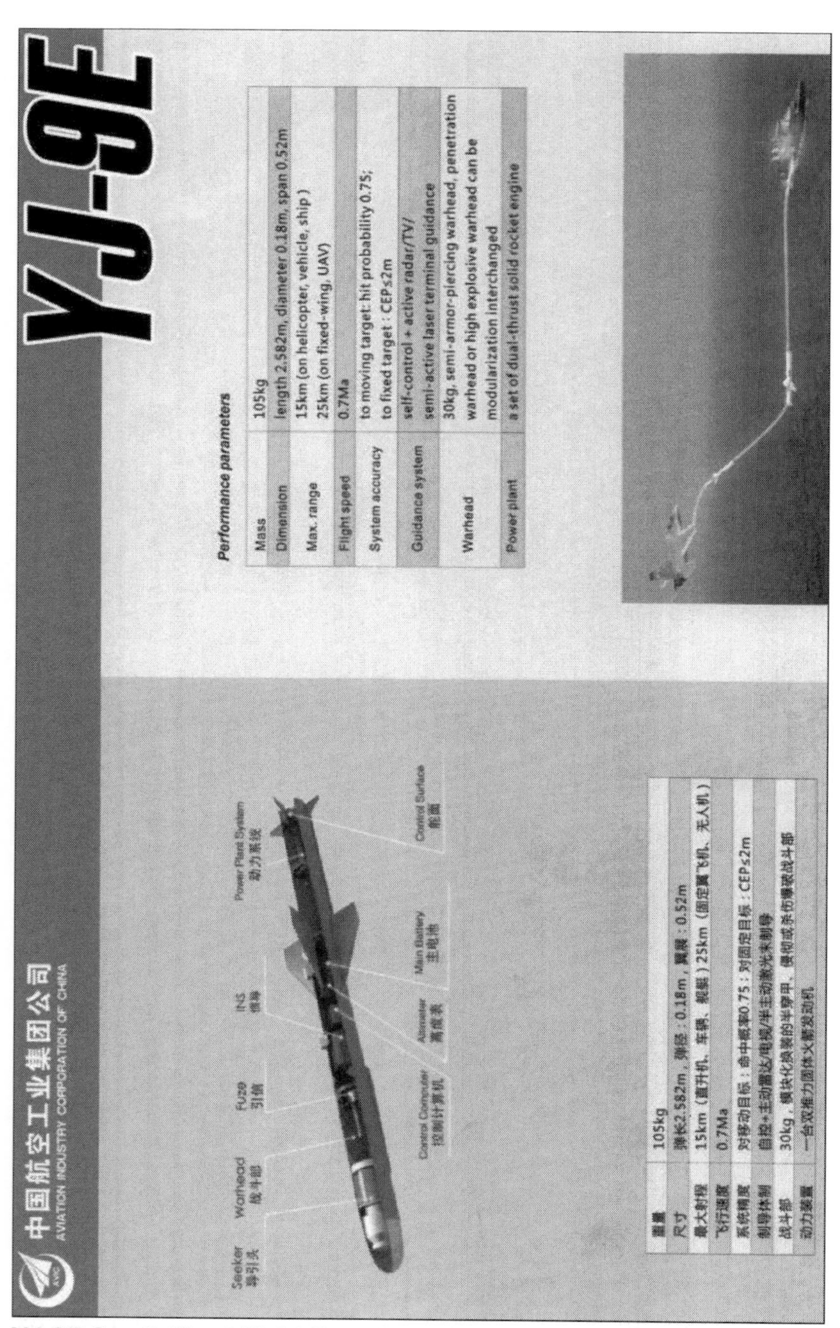

YJ-9E Small-Sized Multi-Function Missile. Aviation Industry Corporation of China (AVIC). 2016 Zhuhai Airshow. (2/5).

Performance parameters	
Mass	105kg
Dimension	length 2.582m, diameter 0.18m, span 0.52m
Max. range	15km (on helicopter, vehicle, ship) 25km (on fixed-wing, UAV)
Flight speed	0.7Ma
System accuracy	to moving target: hit probability 0.75; to fixed target : CEP≤2m
Guidance system	self-control + active radar/TV/ semi-active laser terminal guidance
Warhead	30kg, semi-armor-piercing warhead, penetration warhead or high explosive warhead can be modularization interchanged
Power plant	a set of dual-thrust solid rocket engine

YJ-9E Small-Sized Multi-Function Missile. Aviation Industry Corporation of China (AVIC). 2016 Zhuhai Airshow. (3/5).

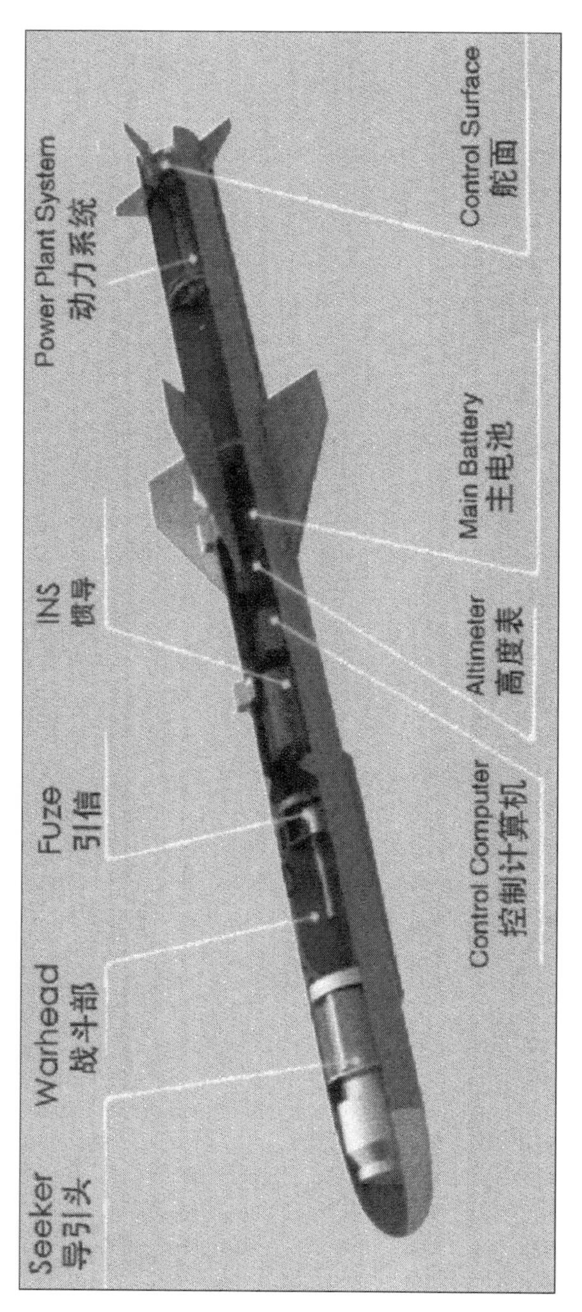

YJ-9E Small-Sized Multi-Function Missile. Aviation Industry Corporation of China (AVIC). 2016 Zhuhai Airshow. (4/5).

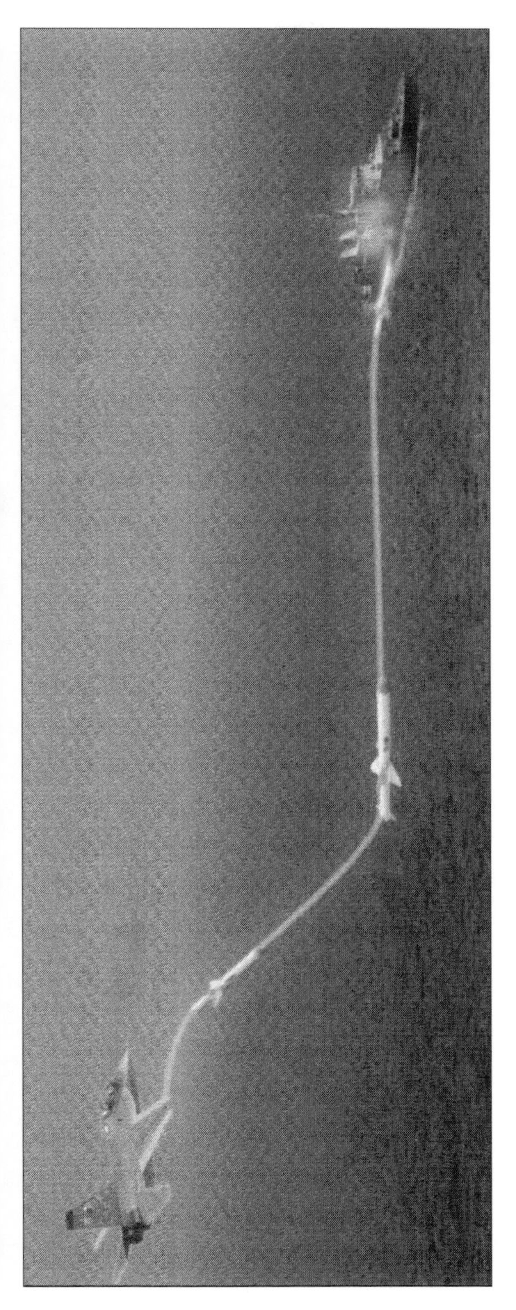

YJ-9E Small-Sized Multi-Function Missile. Aviation Industry Corporation of China (AVIC). 2016 Zhuhai Airshow. (5/5).

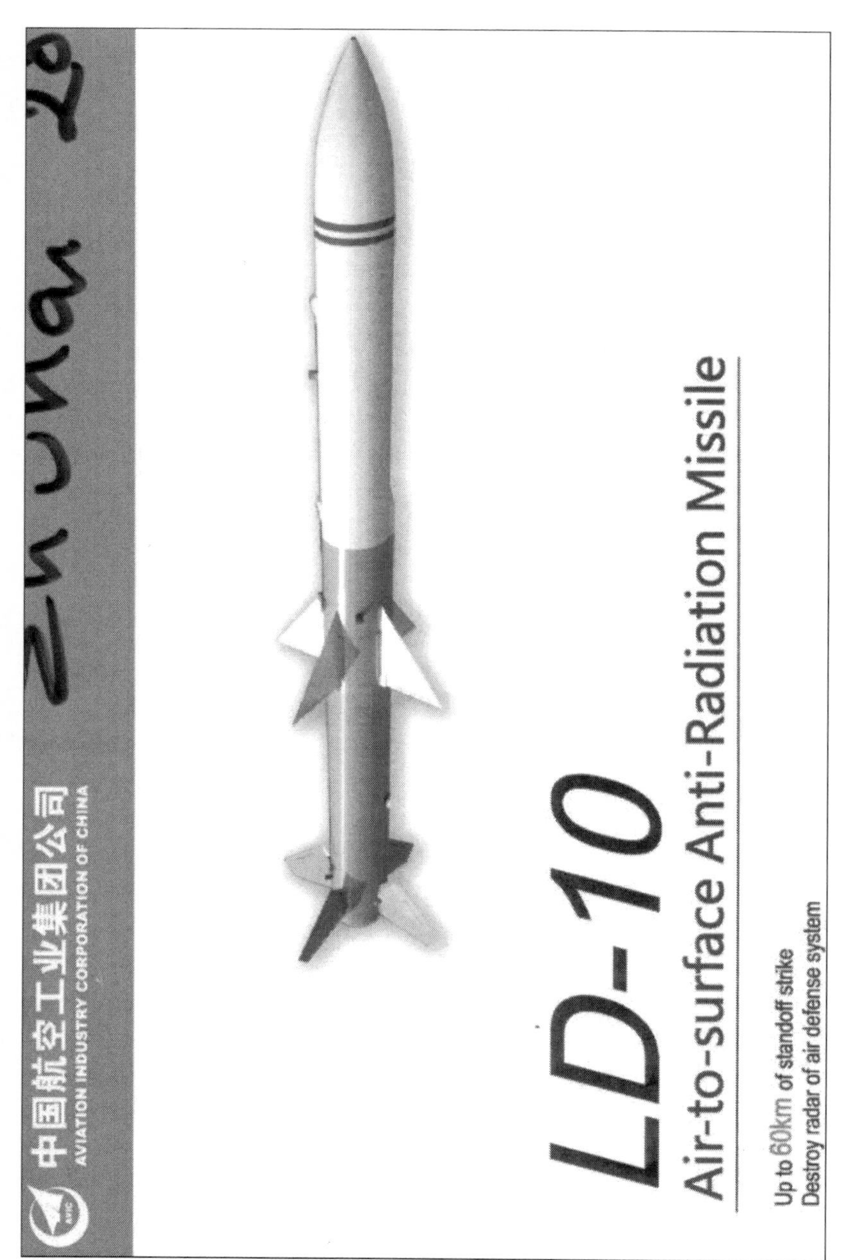

LD-10 (PL-12) Air-to-Surface Anti-Radiation Missile, Aviation Industry Corporation of China (AVIC), 2014 Zhuhai Airshow (1/3).

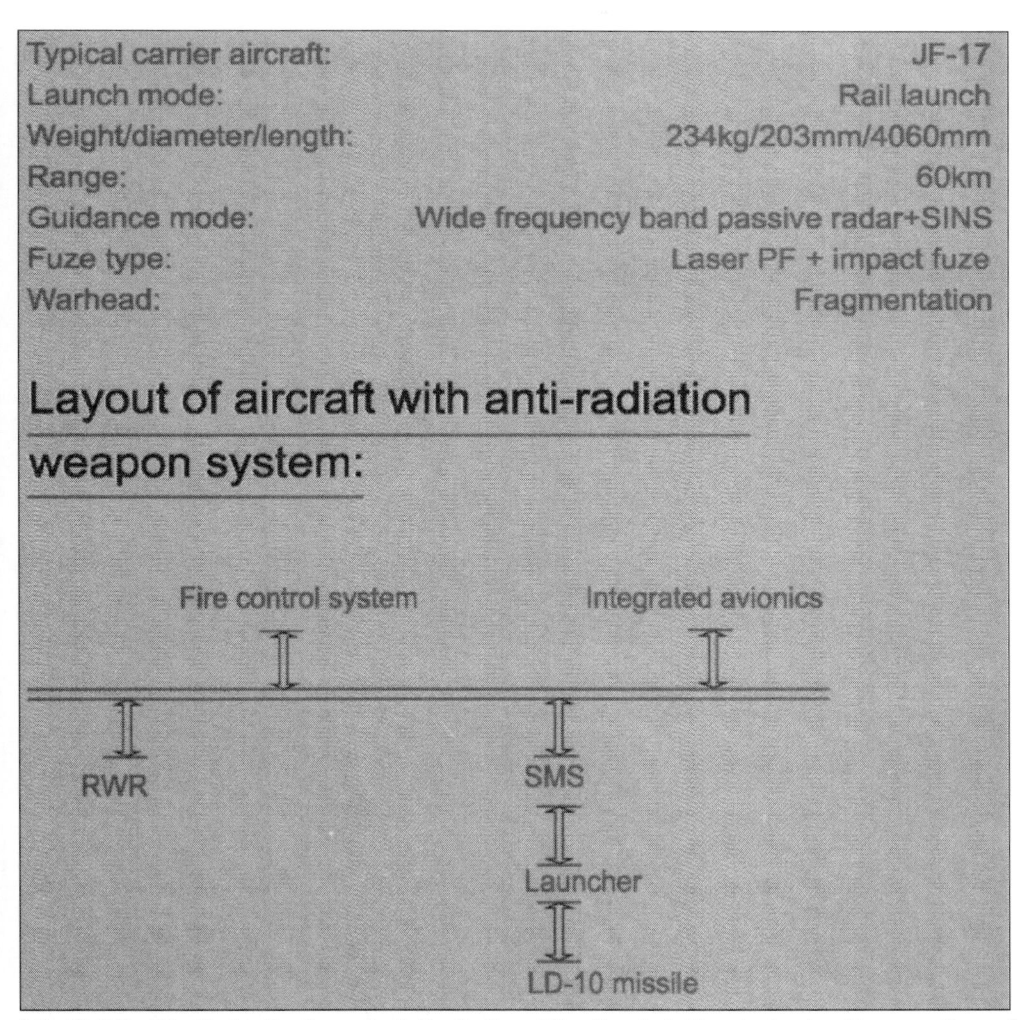

LD-10 (PL-12) Air-to-Surface Anti-Radiation Missile, Aviation Industry Corporation of China (AVIC), 2014 Zhuhai Airshow (2/3).

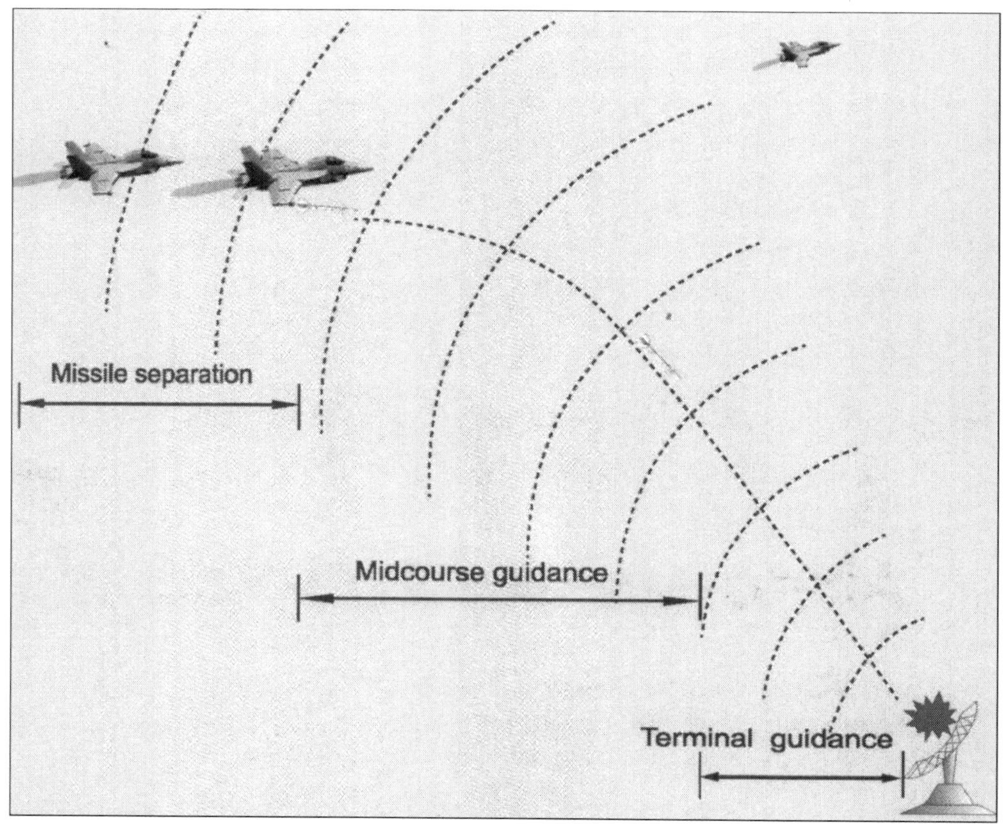

LD-10 (PL-12) Air-to-Surface Anti-Radiation Missile, Aviation Industry Corporation of China (AVIC), 2014 Zhuhai Airshow (3/3).

AIR-TO-AIR MISSILES

PL-5II Advanced Short-Range Air-to-Air Missile, Aviation Industry Corporation of China (AVIC), 2010 Zhuhai Airshow (1/3).

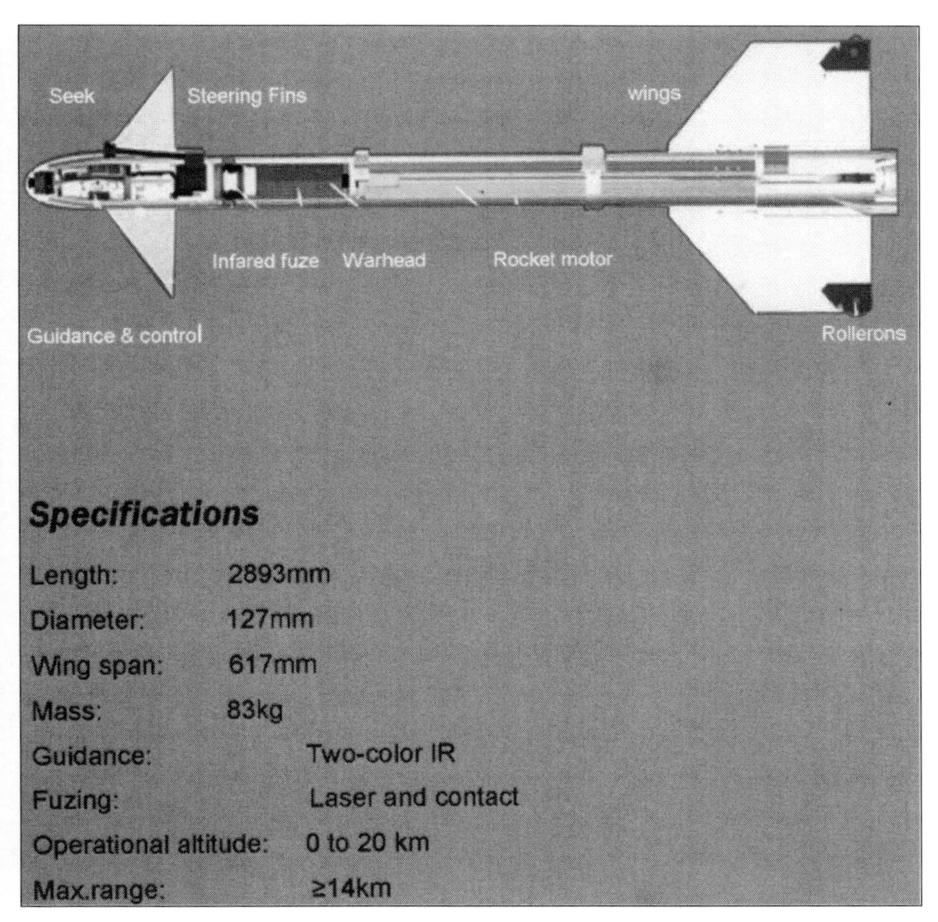

PL-5II Advanced Short-Range Air-to-Air Missile, Aviation Industry Corporation of China (AVIC), 2010 Zhuhai Airshow (2/3).

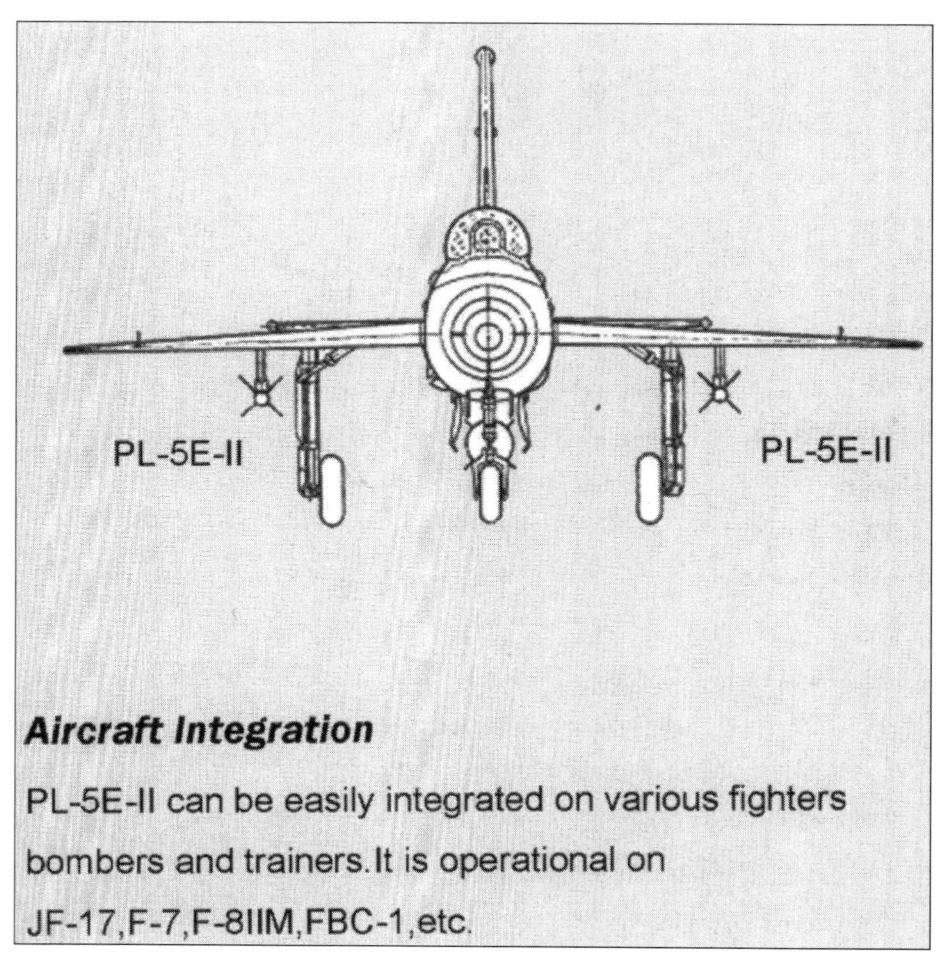

PL-5II Advanced Short-Range Air-to-Air Missile, Aviation Industry Corporation of China (AVIC), 2010 Zhuhai Airshow (3/3).

PL-5II Advanced Short-Range Air-to-Air Missile, China National Aero-Technology Import and Export Corp. (CATIC), 2012 Singapore Airshow (1/3).

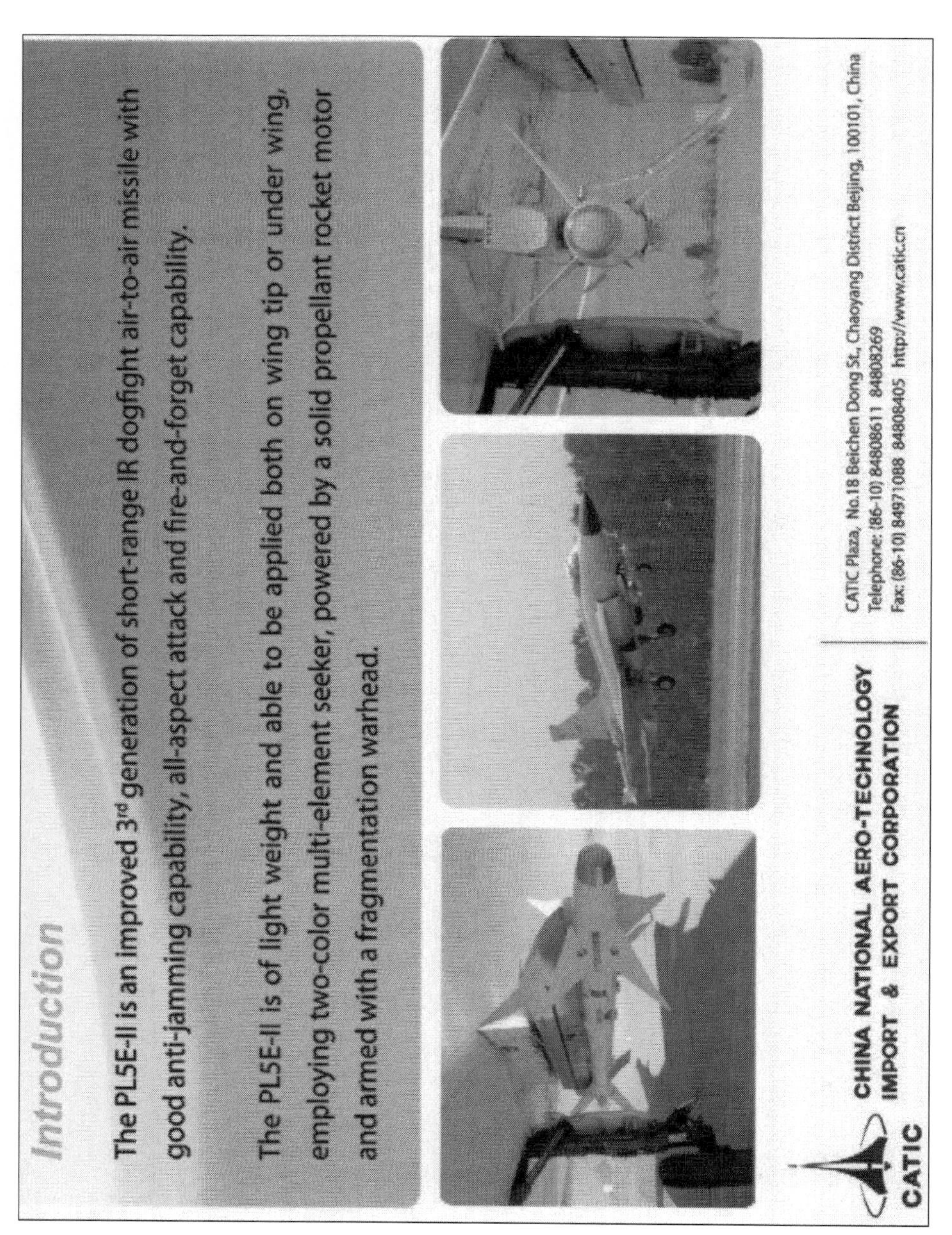

PL-5II Advanced Short-Range Air-to-Air Missile, China National Aero-Technology Import and Export Corp. (CATIC), 2012 Singapore Airshow (2/3).

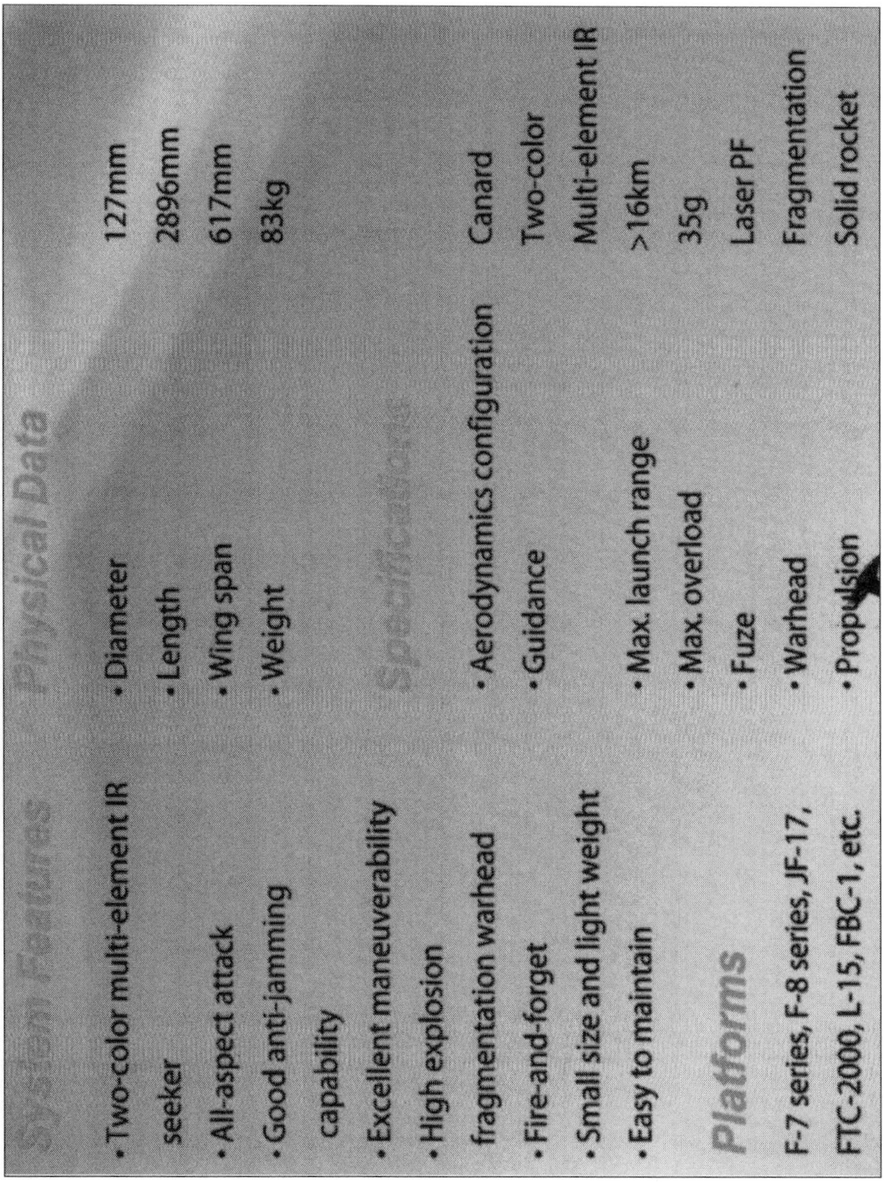

System Features

- Two-color multi-element IR seeker
- All-aspect attack
- Good anti-jamming capability
- Excellent maneuverability
- High explosion fragmentation warhead
- Fire-and-forget
- Small size and light weight
- Easy to maintain

Platforms

F-7 series, F-8 series, JF-17, FTC-2000, L-15, FBC-1, etc.

Physical Data

• Diameter	127mm
• Length	2896mm
• Wing span	617mm
• Weight	83kg

Specifications

• Aerodynamics configuration	Canard
• Guidance	Two-color Multi-element IR
• Max. launch range	>16km
• Max. overload	35g
• Fuze	Laser PF
• Warhead	Fragmentation
• Propulsion	Solid rocket

PL-5II Advanced Short-Range Air-to-Air Missile, China National Aero-Technology Import and Export Corp. (CATIC), 2012 Singapore Airshow (3/3).

PL-9C Short-Range Multi-Mission Missile, Aviation Industry Corporation of China (AVIC), 2014 Zhuhai Airshow (1/3).

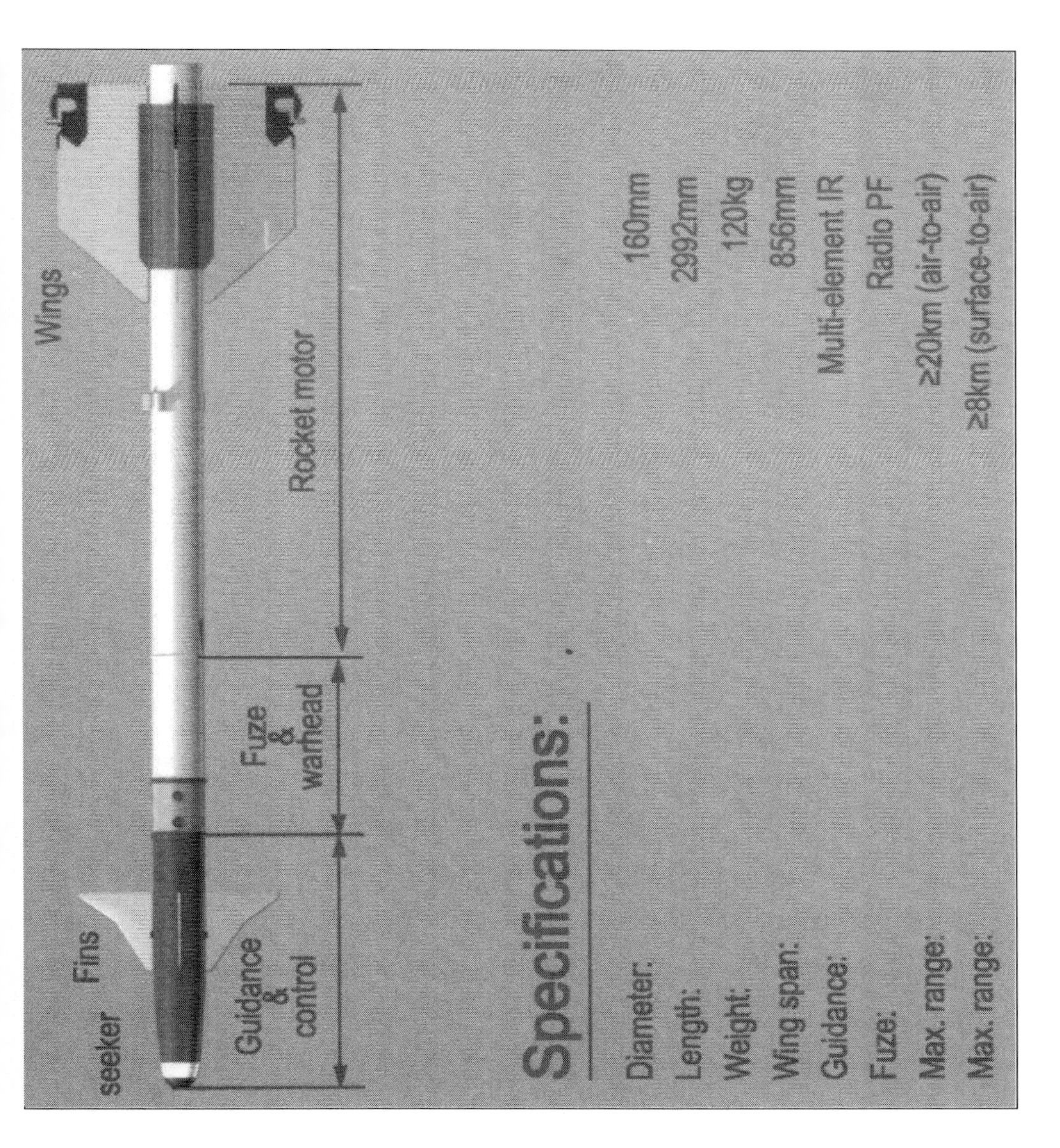

PL-9C Short-Range Multi-Mission Missile, Aviation Industry Corporation of China (AVIC), 2014 Zhuhai Airshow (2/3).

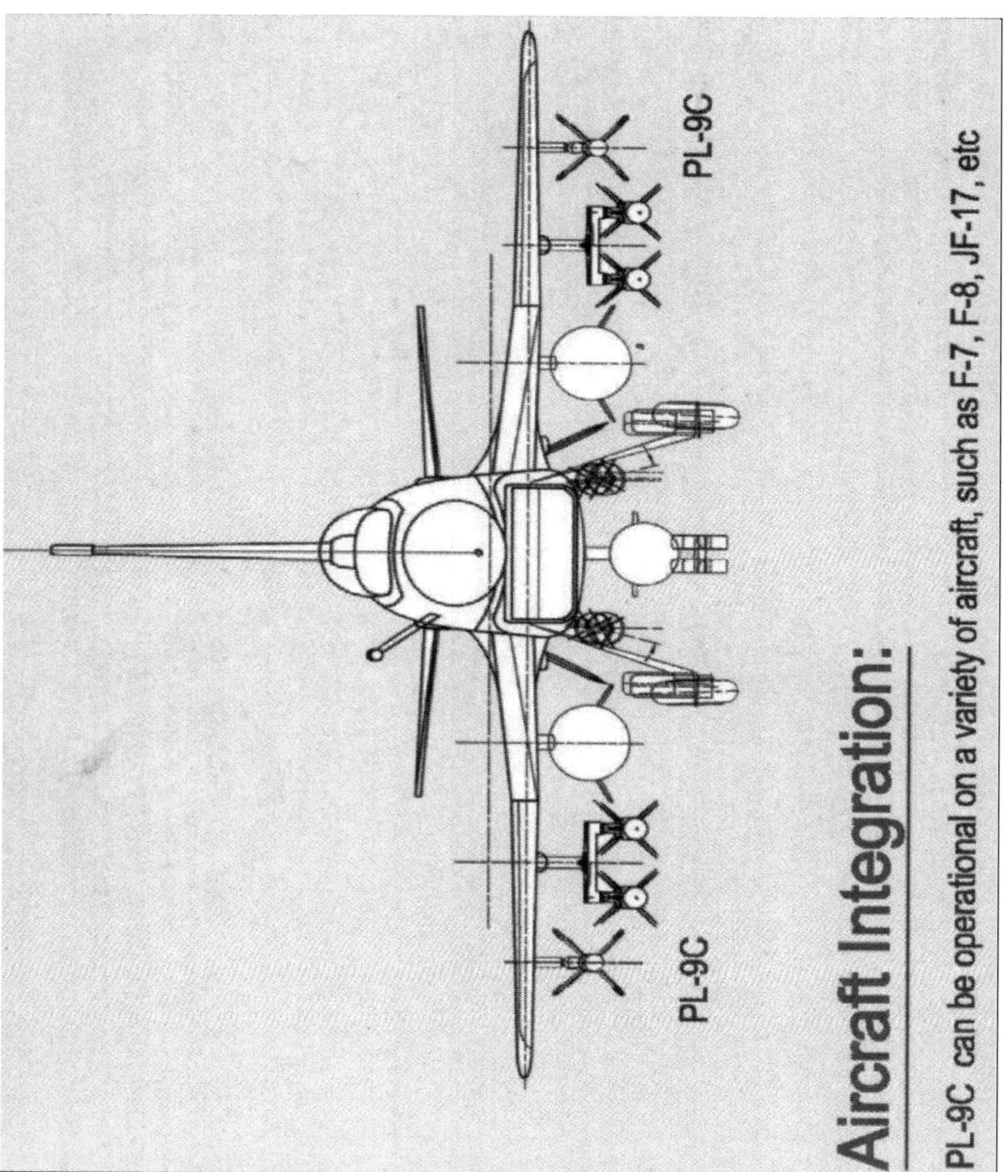

PL-9C Short-Range Multi-Mission Missile, Aviation Industry Corporation of China (AVIC), 2014 Zhuhai Airshow (3/3).

SD-10A Advanced Medium-Range Air-to-Air Missile, China National Aero-Technology Import and Export Corp. (CATIC), 2009 Dubai Airshow, (1/4). Note: two mock-ups at Dubai Airshow.

SD-10A Advanced Medium-Range Air-to-Air Missile, China National Aero-Technology Import and Export Corp. (CATIC), 2009 Dubai Airshow, (2/4). Note: two mock-ups at Dubai Airshow.

- BVR capability
- All-aspect, all-weather attack
- Multi-target capability
- Fire-and-forget
- Anti-jamming capability, countering passive, active or support jamming

Guidance Modes

- Combined mode
- Fire-and-forget mode
- Passive mode

Physical Data

• Length	3934mm
• Diameter	203mm
• Wing span	670mm
• Weight	199kg

Specifications

• Aerodynamic configuration	Conventional
• Operational altitude	0 to 21km
• Max. overload	38g
• Max. launch range	70kg
• Max. speed	> M5
• Fuze	Radio PF
• Warhead	Discrete rod
• Propulsion	Solid rocket

SD-10A Advanced Medium-Range Air-to-Air Missile, China National Aero-Technology Import and Export Corp. (CATIC), 2009 Dubai Airshow, (3/4). Note: two mock-ups at Dubai Airshow.

Introduction

SD-10A is a 4th generation of advanced medium-range intercept missile, which employs the inertial navigation plus data-link in the mid-course and the active radar guidance at the terminal course.

SD-10A features high guidance accuracy, high kill probability, strong anti-jamming capability, and all-weather, all-altitude and all-aspect attack capabilities, as well as BVR launch and fire-and-forget.

SD-10A can be used for not only medium-range interception, but also dogfight.

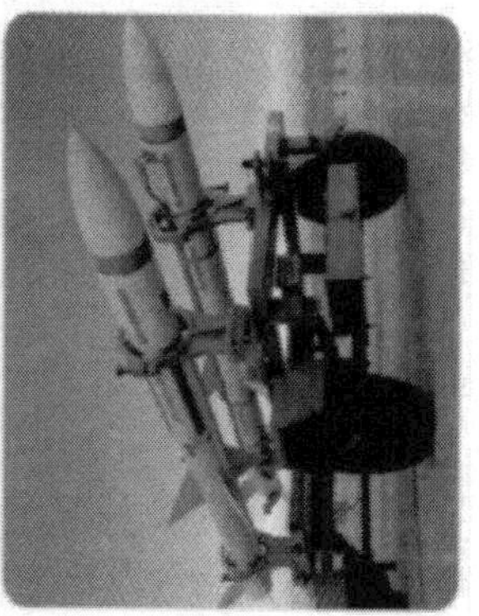

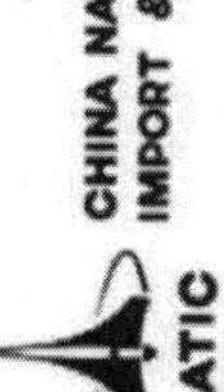

CHINA NATIONAL AERO-TECHNOLOGY
IMPORT & EXPORT CORPORATION

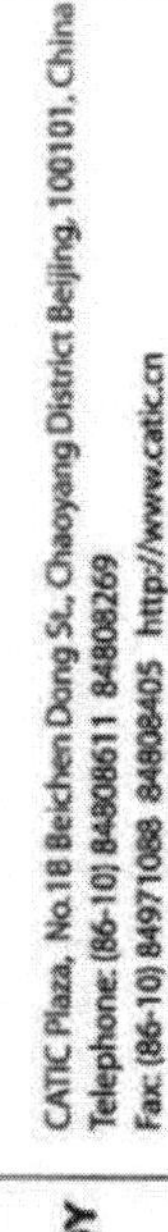

SD-10A Advanced Medium-Range Air-to-Air Missile, China National Aero-Technology Import and Export Corp. (CATIC), 2009 Dubai Airshow, (4/4). Note: two mock-ups at Dubai Airshow.

SD-10A Advanced Medium-Range Air-to-Air Missile. Aviation Industry Corporation of China (AVIC). 2014 Zhuhai Airshow. (1/2).

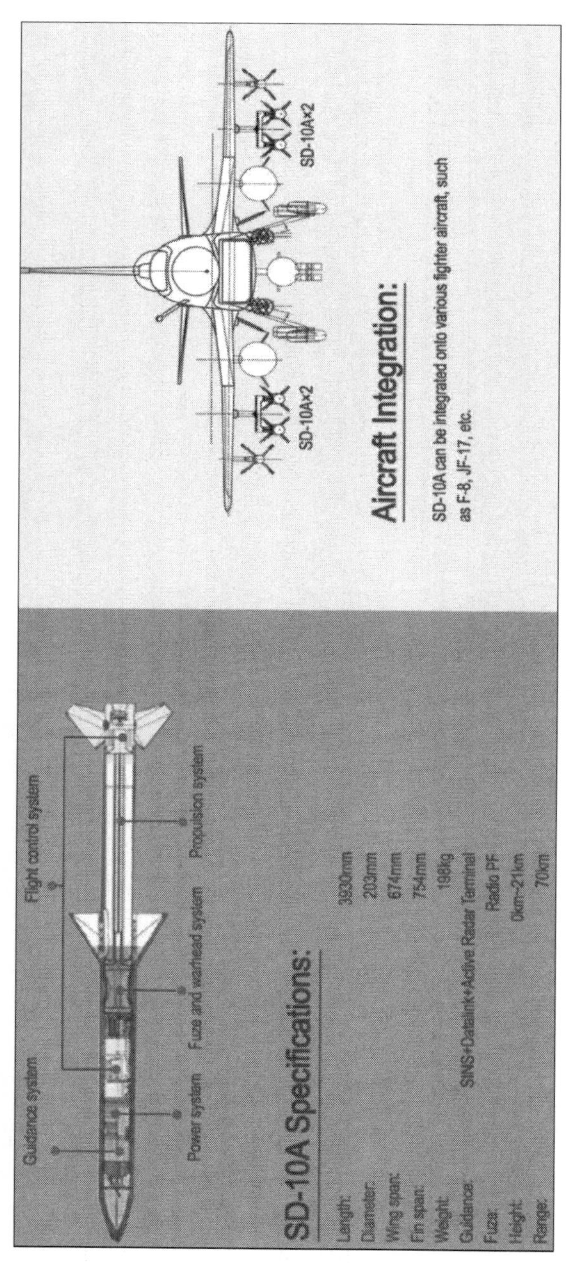

SD-10A Advanced Medium-Range Air-to-Air Missile. Aviation Industry Corporation of China (AVIC). 2014 Zhuhai Airshow. (2/2).

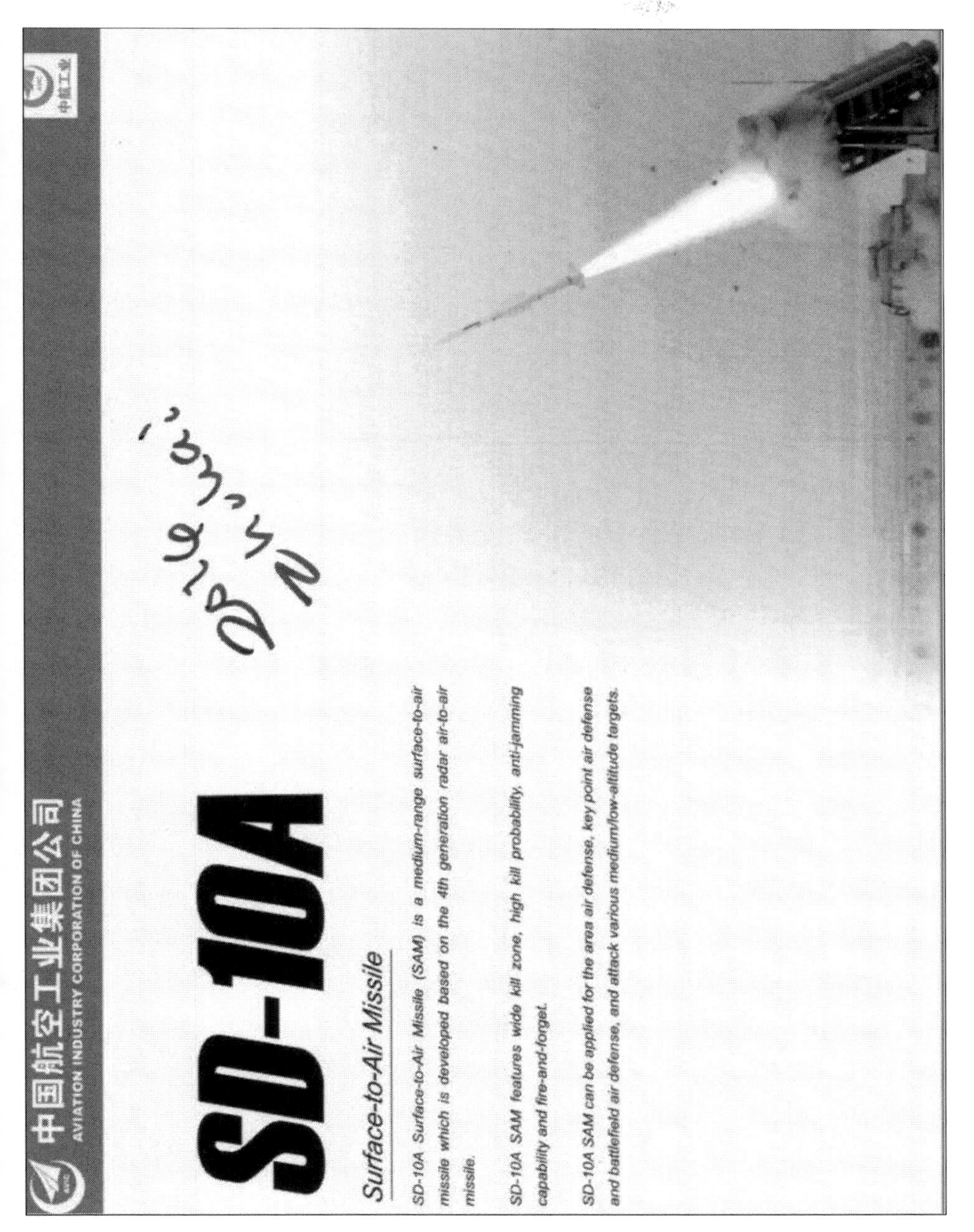

SD-10A Surface-to-Air Missile (variant of the SD-10A Air-to-Air Missile). Aviation Industry Corporation of China (AVIC). 2016 Zhuhai Airshow. (1/2).

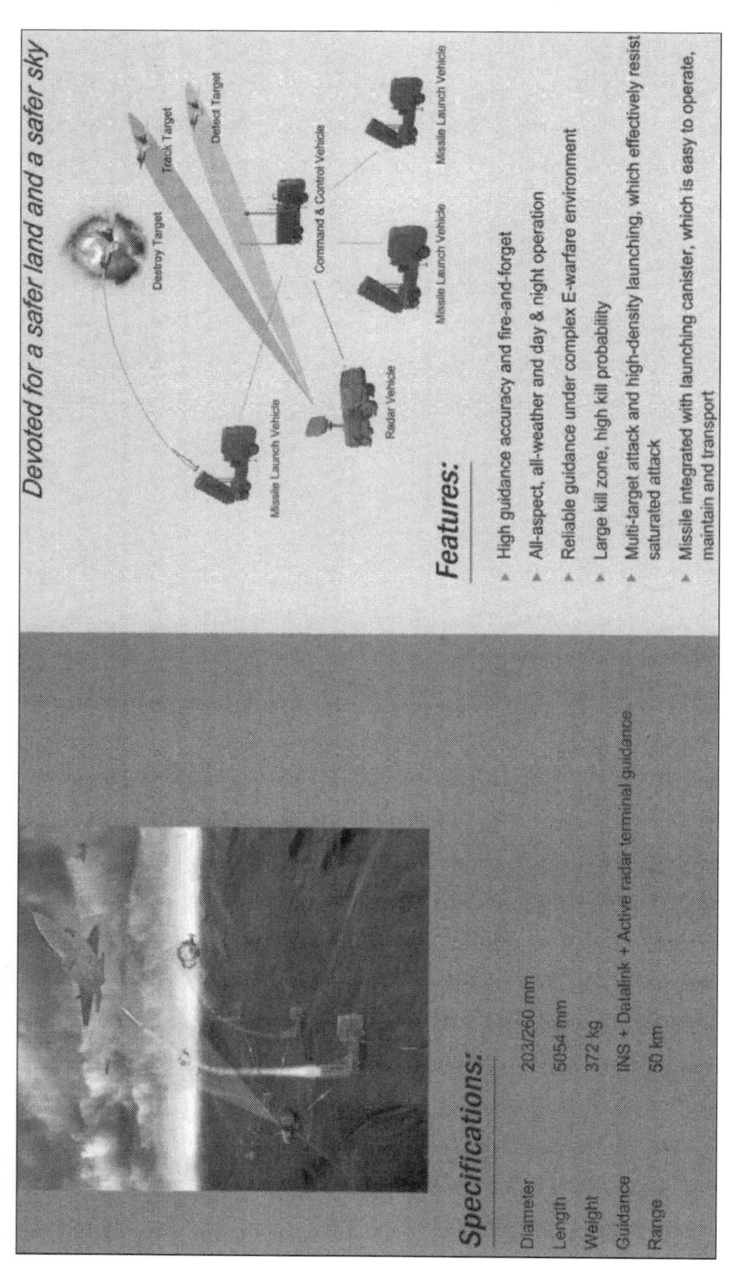

SD-10A Surface-to-Air Missile (variant of the SD-10A Air-to-Air Missile). Aviation Industry Corporation of China (AVIC). 2016 Zhuhai Airshow. (2/2).

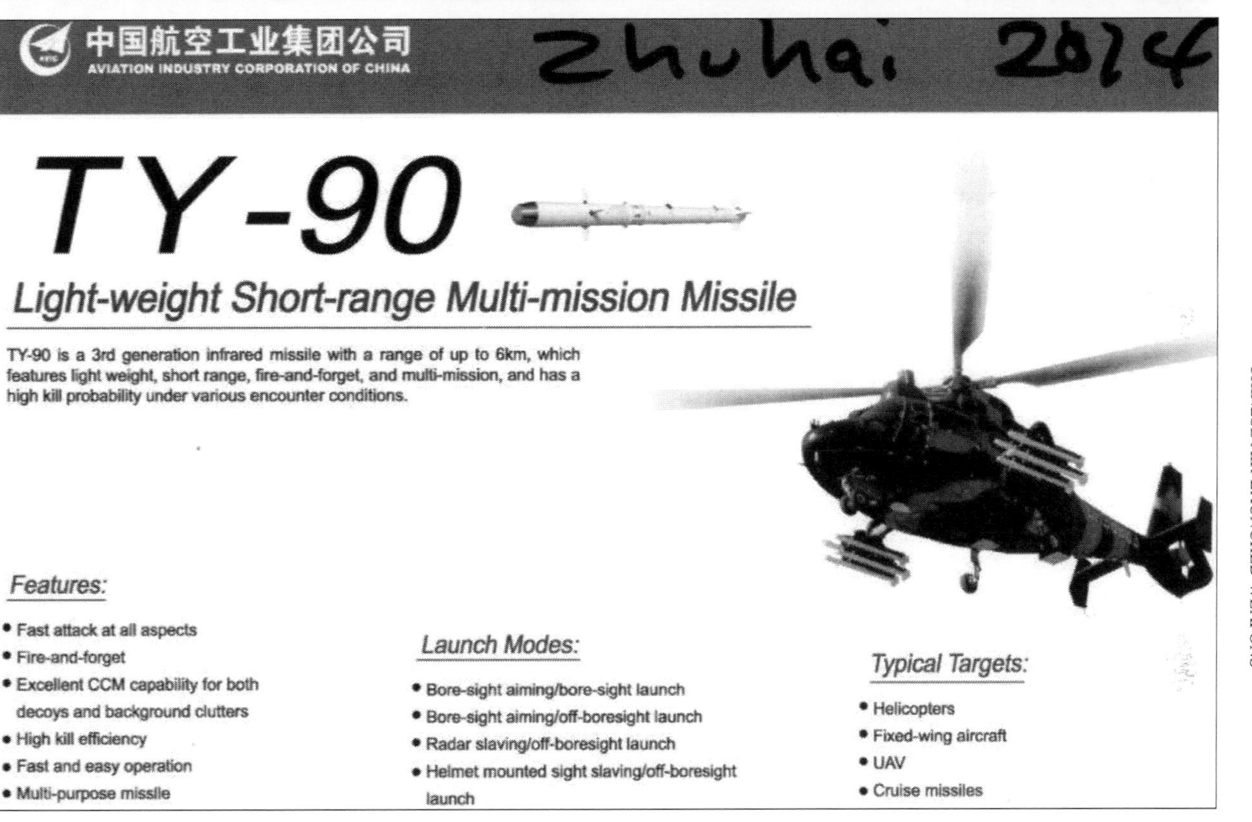

TY-90 Light-Weight Short-Range Multi-Mission Missile. Aviation Industry Corporation of China (AVIC). 2014 Zhuhai Airshow. (1/2).

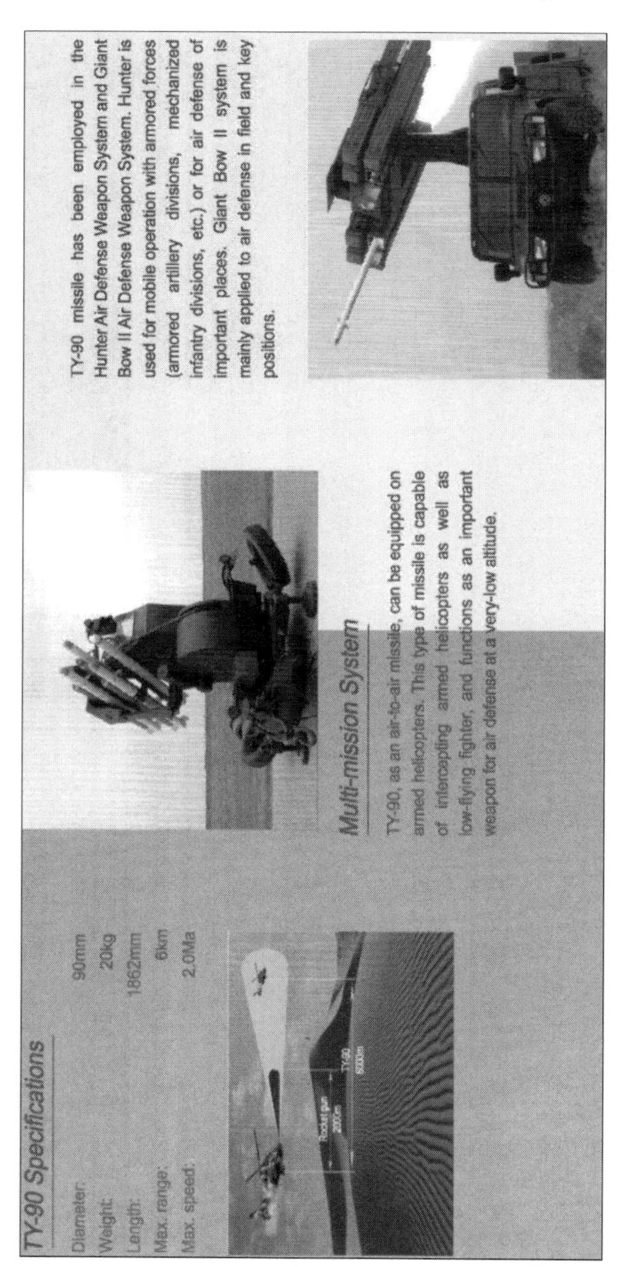

TY-90 Light-Weight Short-Range Multi-Mission Missile. Aviation Industry Corporation of China (AVIC). 2014 Zhuhai Airshow. (2/2).

BOMBS

250-3 Low-Drag Aviation Demolition & Fragmentation Bomb

250-3 low-drag aviation demolition & fragmentation bomb major strikes light armor infantry vehicles, armor carrier vehicles, auto cars, trains, radars, artillery, oil plant, open air fuel magazine, depot, small scale ships and personnel, etc. This product has wonderful aerodynamic shape, stable ballistic flying, enabling the bomb to be carried on the high speed aircraft for operational using; this bomb can produce even, intensive, high speed fragmentations after exploding, this bomb is a newly aviation bomb of big power, high performance.

China South Industries Group Corporation 0731-86137326

250-3 Low-Drag Aviation Demolition and Fragmentation Bomb. China South Industries Group Corp. 2012 Zhuhai Airshow (1/1).

250-4 Low-Drag, Low-Altitude Aviation Demolition & Fragmentation Bomb

250-4 low-drag, low-altitude aviation demolition & fragmentation bomb major strikes light armor infantry vehicles, armor carrier vehicles, auto cars, trains, radars, artillery, oil plant, open air fuel magazine, depot, small scale ships and personnel, etc. This product has wonderful aerodynamic shape, stable ballistic flying, enabling the bomb to be carried on the high speed aircraft for operational using; this product adapts flexible drag parachute, so it can use in the condition of low-altitude and extreme low-altitude. This bomb can produce even, intensive, high speed fragmentations after exploding, this bomb is a newly aviation bomb of big power, high performance.

China South Industries Group Corporation 0731-86137326

250-4 Low-Drag, Low-Altitude Aviation Demolition and Fragmentation Bomb. China South Industries Group Corp. 2012 Zhuhai Airshow (1/1).

CS/BBC5 Type 500kg Guided & Gliding Cluster Bomb

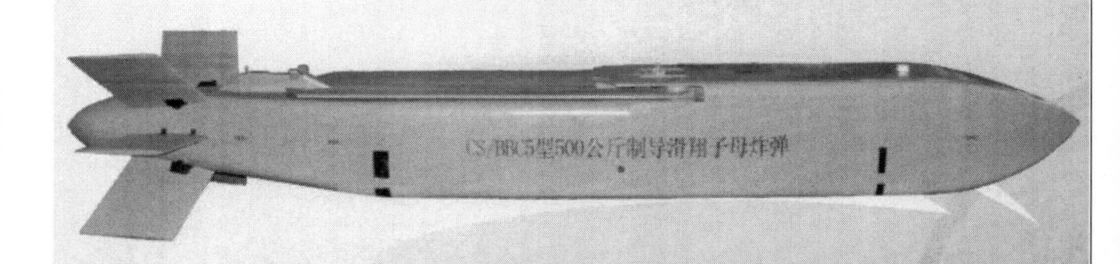

CS/BBC5 type 500kg guided & gliding cluster bomb is one kind of standoff attack weapon of short range, having the functions of low cost, long range, high power, multipurpose use, all-weather use and fire after forget etc, could drop from outside of the enemy ground firing area, effectively attack the enemy's high-value targets, to win air supremacy, weaken the enemy's war capabilities and potential. Could bring multiple damage effect, and attack all kinds' ground targets through loading different submunition; the product also could load sole warhead to realize crucial role.

China South Industries Group Corporation 0731-86137326

CS/BBC5 Type 500kg Guided and Gliding Cluster Bomb. China South Industries Group Corp. 2012 Zhuhai Airshow (1/1).

CS/BBF1 Type 250kg Fuel Air Explosive Bomb

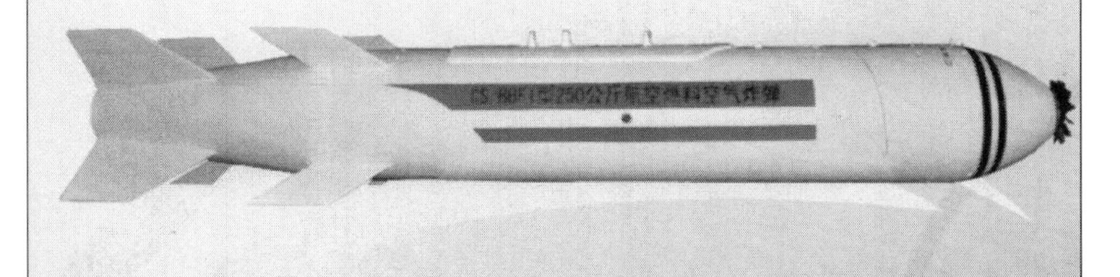

CS/BBF1 type 250kg fuel air explosive bomb is a kind of air toward ground area damage weapon, major strikes the ground or water surface all kinds soft and semi-hard targets which sensitize overpressure function of shock wave, such as open-air airplane, bomb launching position, ground radar facilities, gather or march forward motorcade, land mine field, effective strength in the fortification or bunkers, small scale ship, etc. Its damage function mainly is using second detonate technology, forming cloud and mist blasting, producing detonation wave and overpressure of shock wave, at the same time the product can bring burning and heat energy damage. The product has high energy density, damage area amplitude, and multiple damaged factor of characteristic.

China South Industries Group Corporation 0731-86137326

CS/BBF1 Type 250kg Fuel Air Explosive Bomb. China South Industries Group Corp. 2012 Zhuhai Airshow (1/1).

FT-1 Precision Guided Bomb Weapon System. Unknown airshow. China National Precision Machinery Import and Export Corporation (CPMIEC). (1/3).

FT-1 GBU WEAPON SYSTEM

The FT-1 Precision Guided Bomb is a kind of Guided Bomb Unit (GBU) integrating GPS/INS with general low-resistance aviation bomb. Controlled by air rudder at the tail section, the FT-1 possesses the characteristics of low cost and high convenience of the general aviation bombs, as well as the advantage of higher accuracy, longer range, automation and all-weather operation. It greatly improves the delivery precision and survivability of the carrier with the outstanding operational cost-effectiveness.

The FT-1 is applicable to any carrier with airborne inertial guidance system. It can be delivered at high speed and high attitude to precisely attack the surface targets, such as hostile political targets, military headquarters, military plants, harbors, power plants, transformer substations, radar stations, communication centers, air defence establishments and ground forces.

1. Composition

☑ FT-1 Precision Guided Bomb
☑ Ground equipment

2. Main Specifications

☑ Warhead: 500kg low-resistance aviation bomb as the basic type, replaceable by entire penetrative explosion, explosive fragmental cluster, tempera ture pressure, carbon fiber or mine-laying warhead.

☑ CEP: 30m (under the mode of GPS/INS integrated guidance)
☑ Altitude: 5000m~12000m at a speed of 720km/h~1000km/h
☑ Delivery range: 7 ~ 18km
☑ Preparation time for airborne delivery : no more than 5min.
☑ Delivery mode: by ejection, simultaneously or separately
☑ Reliability
a. Captive flight reliability: 0.90
b. Releasing reliability: 0.90
c. Self-control flight reliability:0.88

3. Environmental conditions

☑ Operation environment conditions
a. Temperature: -20ºC ~ +50ºC
b. Relative humidity: ≯95% (25ºC)
c. Relative wind speed: Max. allowable wind speed for delivery is 70m/s at high altitude.
☑ Shelf life: 10 years in central depot at temperature +5ºC ~+30ºC and relative humidity: ≯75% (25ºC)

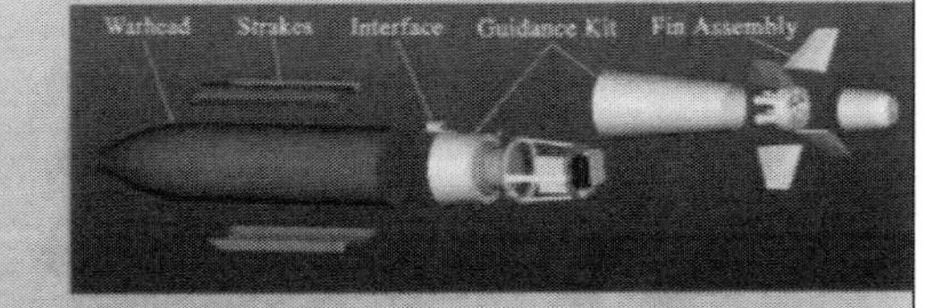

CPMIEC

CHINA NATIONAL PRECISION MACHINERY IMPORT & EXPORT CORPORATION
ADD:NO.30, Haidian Nanlu,Beijing,100080,China
TEL:(8610)68748877 68748894 68748891 FAX:(8610)68748844 http://www.cpmiec.com.cn

FT-1 Precision Guided Bomb Weapon System. Unknown airshow. China National Precision Machinery Import and Export Corporation (CPMIEC). (2/3).

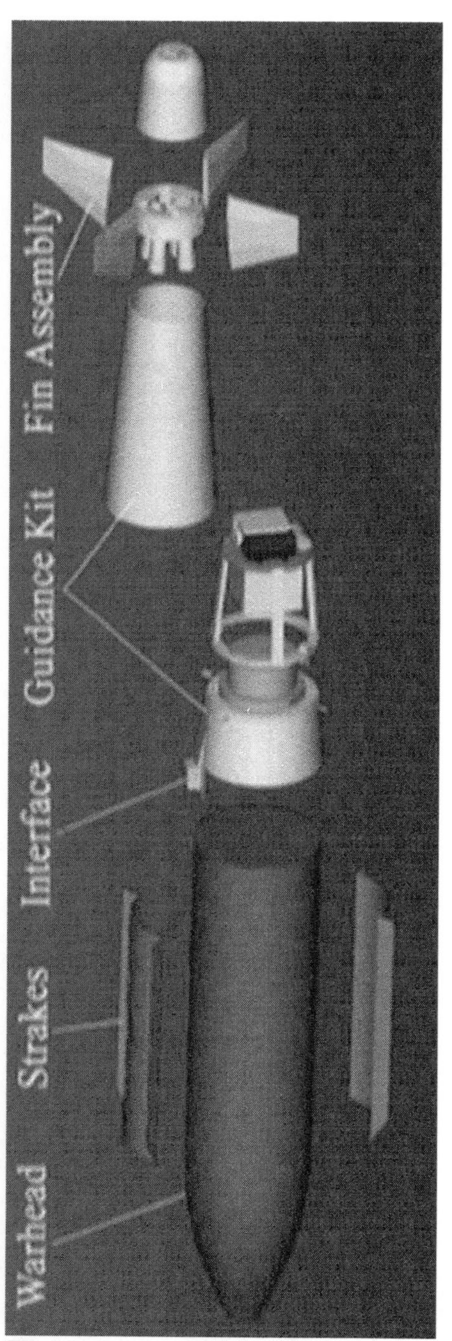

FT-1 Precision Guided Bomb Weapon System. Unknown airshow. China National Precision Machinery Import and Export Corporation (CPMIEC). (3/3).

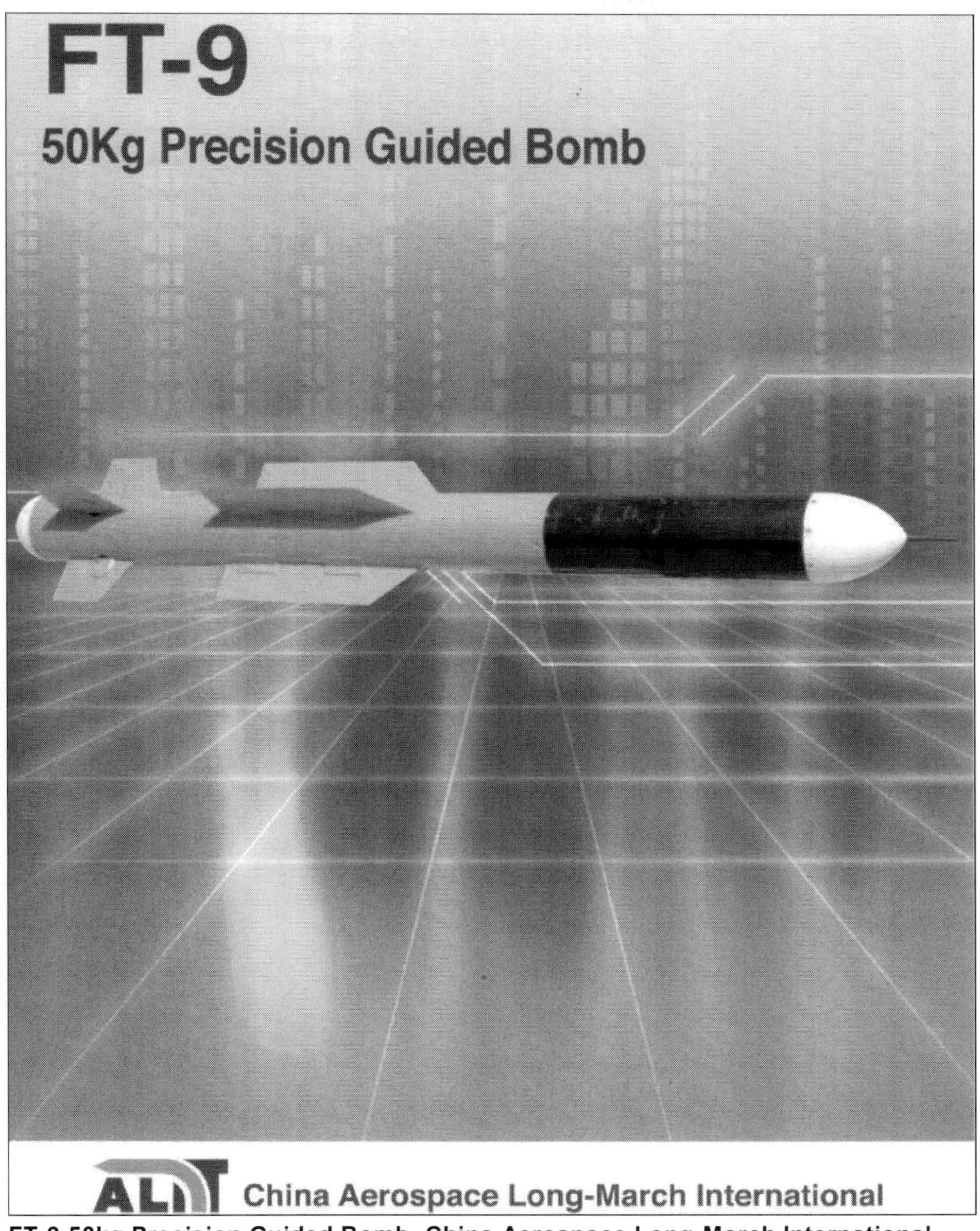

FT-9 50kg Precision Guided Bomb. China Aerospace Long-March International (ALIT). 2014 Zhuhai Airshow (1/2).

FT-9 is a kind of air-to-ground precision guided bomb developed for bombing weak and strong targets, such as armored personnel carrier, missile launching device, air-defense system, personnel, etc. FT-9 possesses many advantages, such as small, light, high accuracy, low-cost. It is developed for operational requirements of high striking precision, small collateral damage and economic affordability and represents the development trend of precision guided bomb.
FT-9 assembles with small diameter blast/fragment warhead. It can be carried by UAV or manned aircraft.

System Composition

FT-9 consists of warhead, fuze, guidance and control cabin, and can apply TV/IR or semi-activate laser seeker. The main electronic devices are inside the guidance and control cabin, including flight controller, inertial measurement unit (IMU), GPS unit, rudder and thermal battery, etc. FT-9 uses blast/fragment warhead and radio proximity /contact electronic fuze.

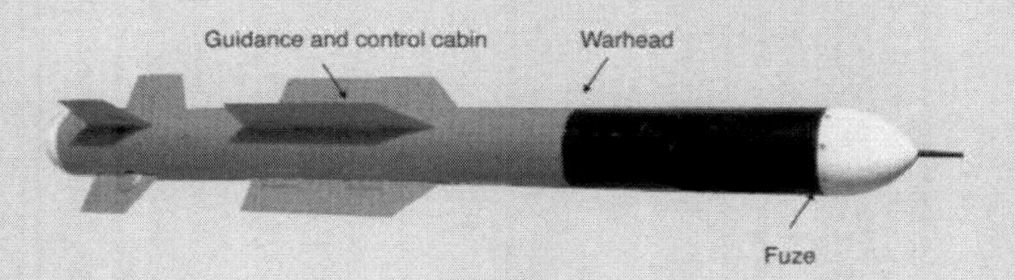

Technical Specifications

Releasing altitude (km)	3 ~ 5 (UAV); 6~10 (Manned aircraft)
Releasing velocity (km/h)	200~260 (UAV); 600~1,000 (Manned aircraft)
Max. range (km)	20 (UAV); 90 (Manned aircraft)
Accuracy (m)	CEP ≤ 15
Guidance mode	INS/GPS
Attack mode	Fixed coordinate
Allowance sorties	10
Accumulated power time (h)	≥ 200
Weight of mass (kg)	≤ 130
Weight of warhead (kg)	≤ 81
Type of warhead	Blast warhead

Add: No.7 building, section 15, No.188 Nansihuan Xilu, Fengtai District, Beijing, P.R. China
Http://www.alitchina.com | Fax: 86-010-56533777 | Tel: 86-010-56533700 | Post Code: 100070

FT-9 50kg Precision Guided Bomb. China Aerospace Long-March International (ALIT). 2014 Zhuhai Airshow (2/2).

LS-6 500kg Precision Guided Bombs. China National Aero-Technology Import and Export Corp. 2008 Zhuhai Airshow (1/4).

Introduction

LS-6 is a 500kg class precision guided bombs, which can carry out standoff strike on fixed ground targets, such as the airports, the seaports, the bridges, the commander centers, etc.

With a range extension unit and a tail guidance unit, a conventional low-drag aerial bomb is cost-effectively re-built to be provided with standoff precision strike capability.

System Features

- All-weather, day & night strike
- Fire and forget
- Anti-jamming capability
- Modular guidance and control unit
- Effective cost

CATIC

CHINA NATIONAL AERO-TECHNOLOGY IMPORT & EXPORT CORPORATION

CATIC Plaza, No.18 Beichen Dong St., Chaoyang District Beijing, 100101, China
Telephone: (86-10) 84808611 84808269
Fax: (86-10) 84971088 84808405 http://www.catic.cn

LS-6 500kg Precision Guided Bombs. China National Aero-Technology Import and Export Corp. (CATIC). 2008 Zhuhai Airshow (2/4).

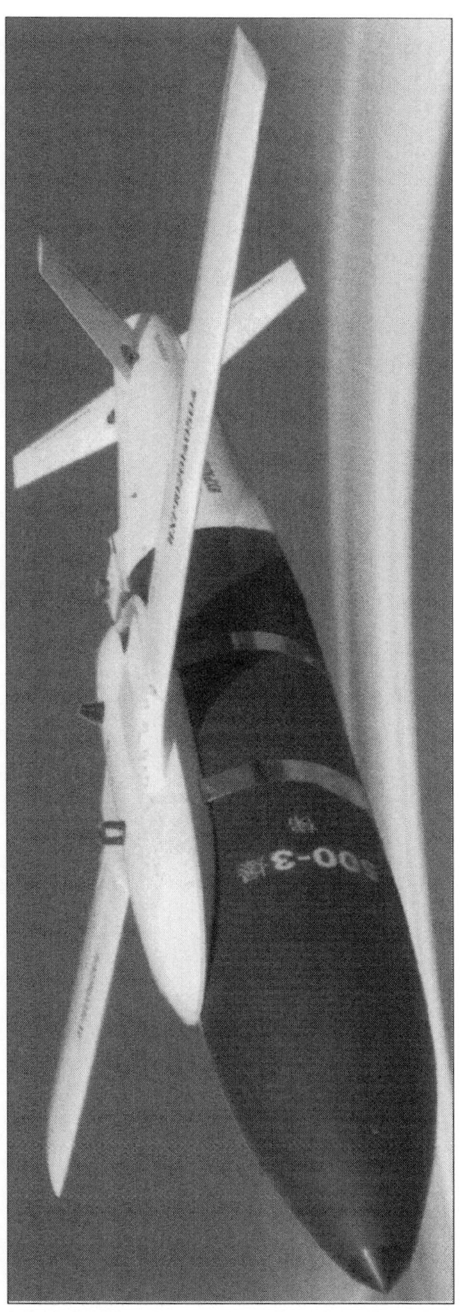

LS-6 500kg Precision Guided Bombs. China National Aero-Technology Import and Export Corp. (CATIC). 2008 Zhuhai Airshow (3/4).

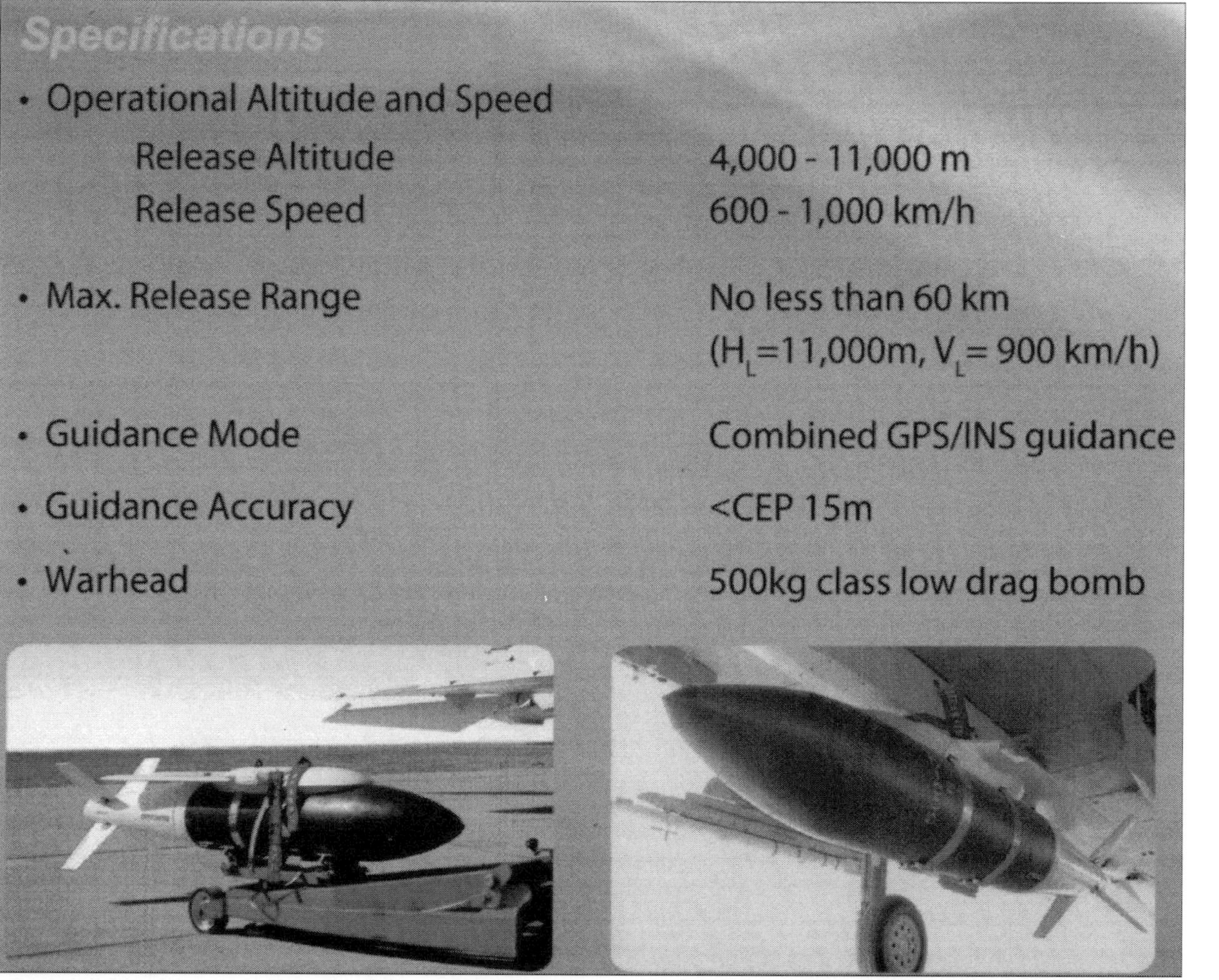

LS-6 500kg Precision Guided Bombs. China National Aero-Technology Import and Export Corp. (CATIC). 2008 Zhuhai Airshow (4/4).

LS-6 GGB/MGB. Aviation Industry Corporation of China (AVIC). 2014 Zhuhai Airshow (1/2)

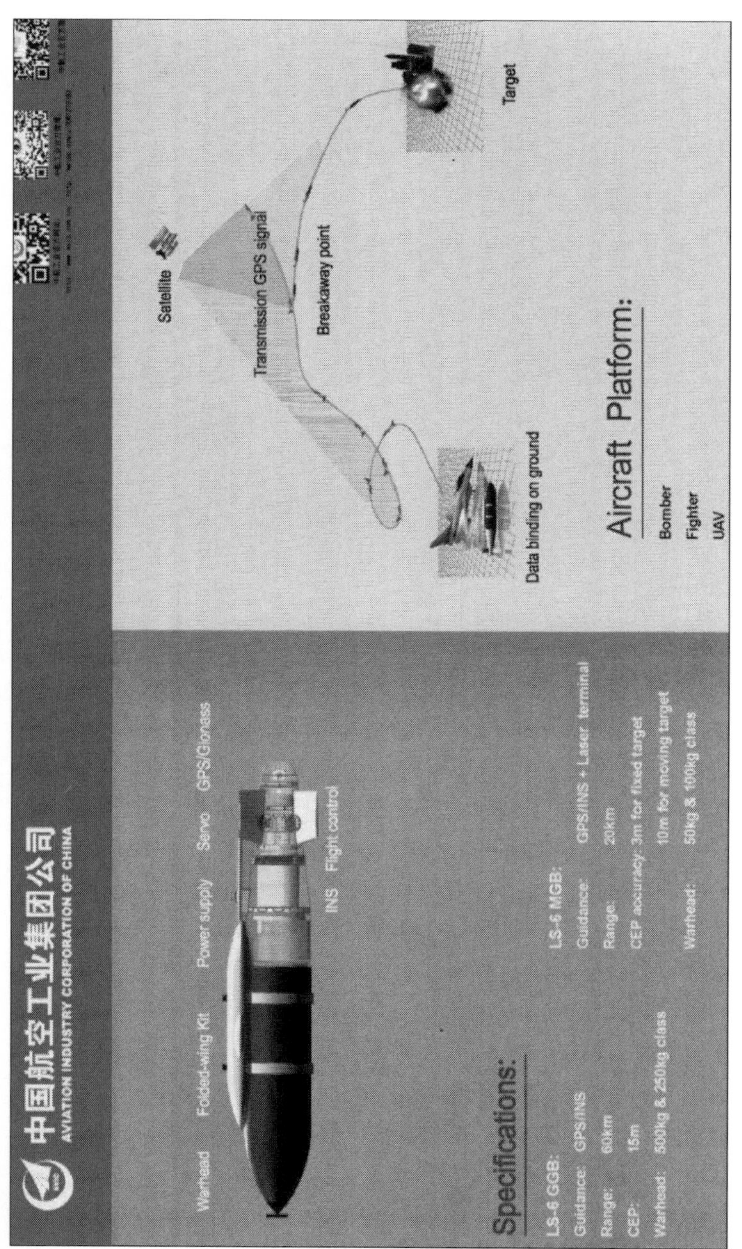

LS-6 GGB/MGB. Aviation Industry Corporation of China (AVIC). 2014 Zhuhai Airshow (2/2).

LS-6 Miniature Guided Bomb. Aviation Industry Corporation of China (AVIC), 2016 Zhuhai Airshow (1/2).

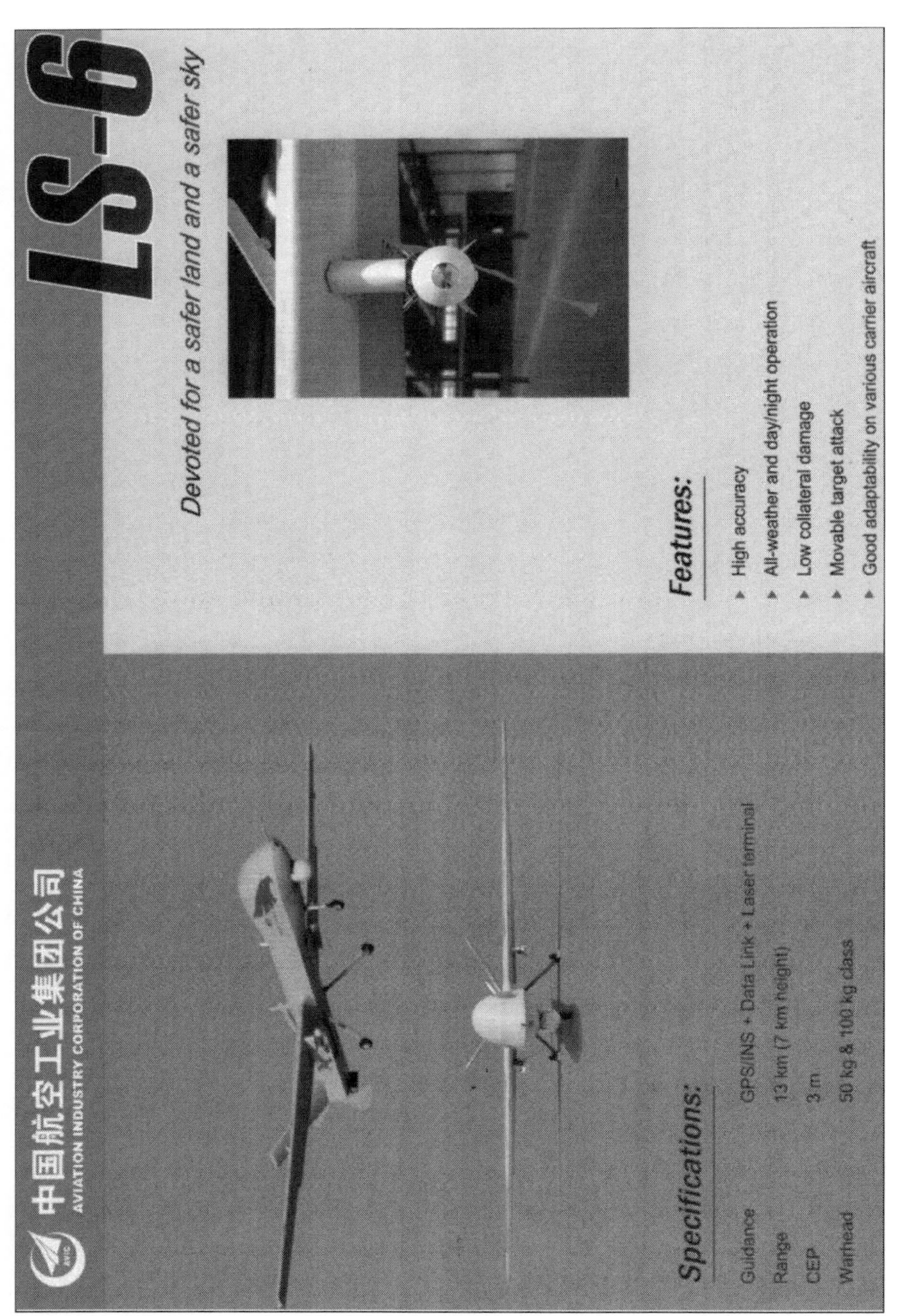

LS-6 Miniature Guided Bomb. Aviation Industry Corporation of China (AVIC), 2016 Zhuhai Airshow (2/2).

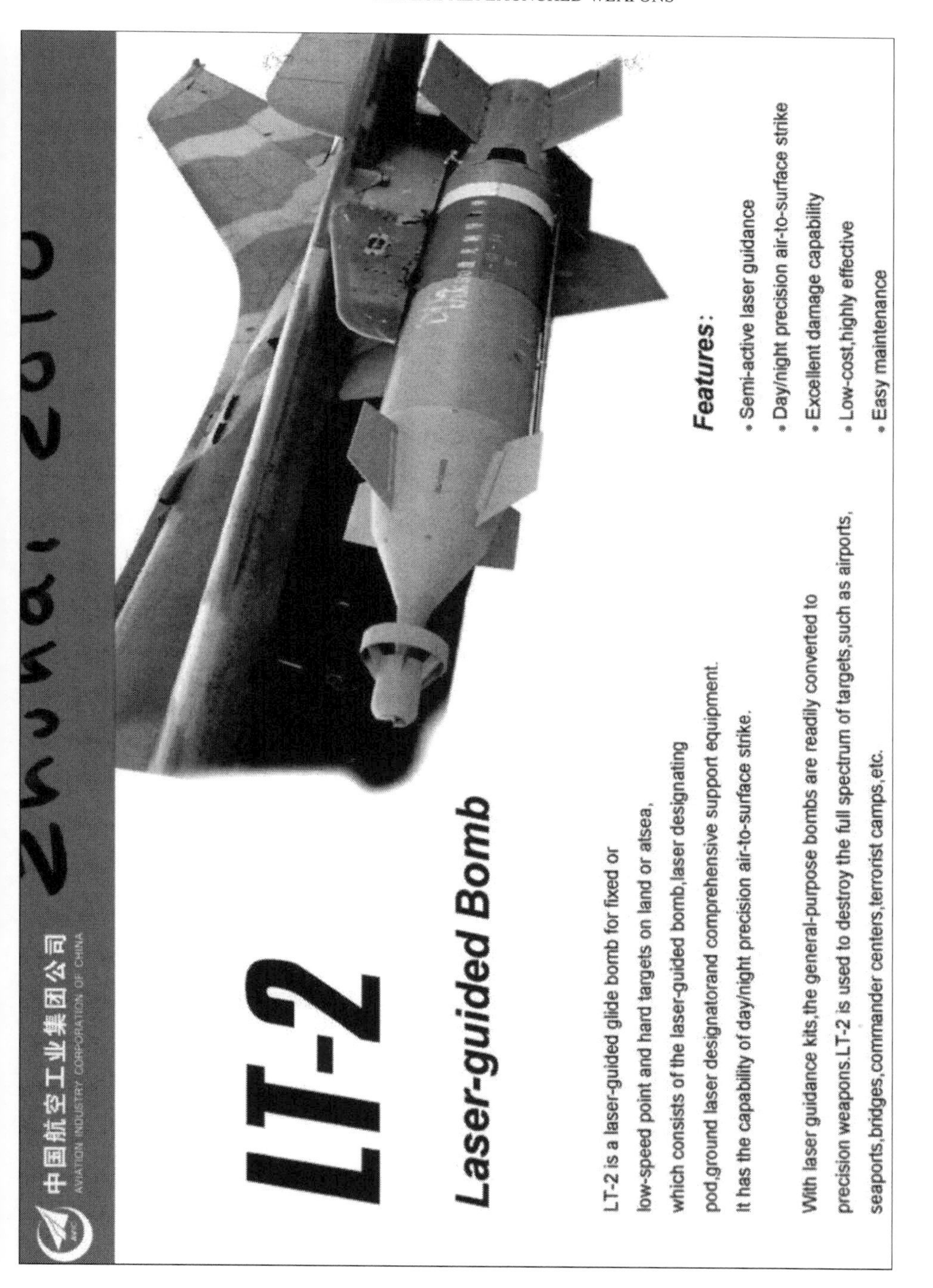

LT-2 Laser Guided Bomb. Aviation Industry Corporation of China (AVIC). 2010 Zhuhai Airshow (1/2).

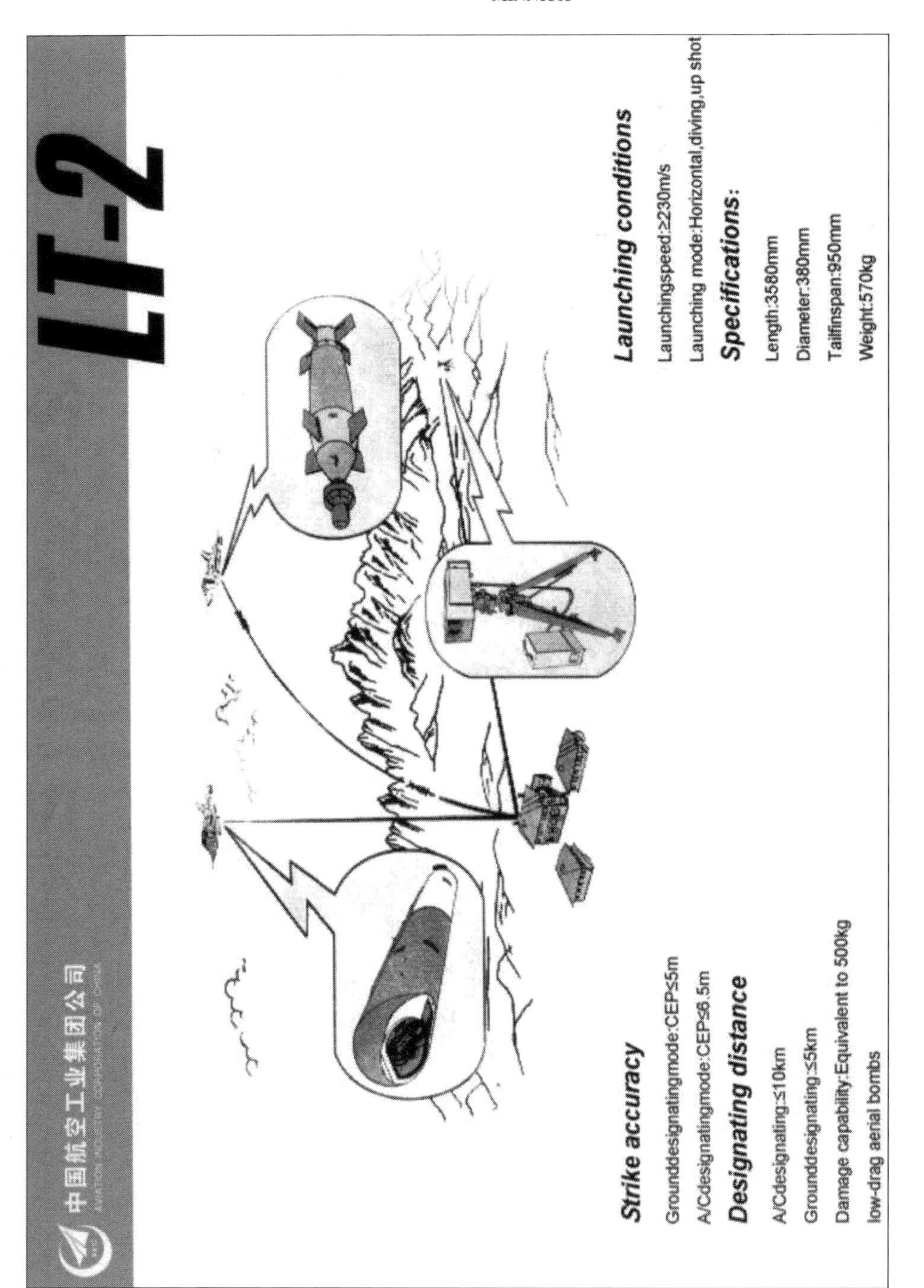

LT-2 Laser Guided Bomb. Aviation Industry Corporation of China (AVIC). 2010 Zhuhai Airshow (2/2).

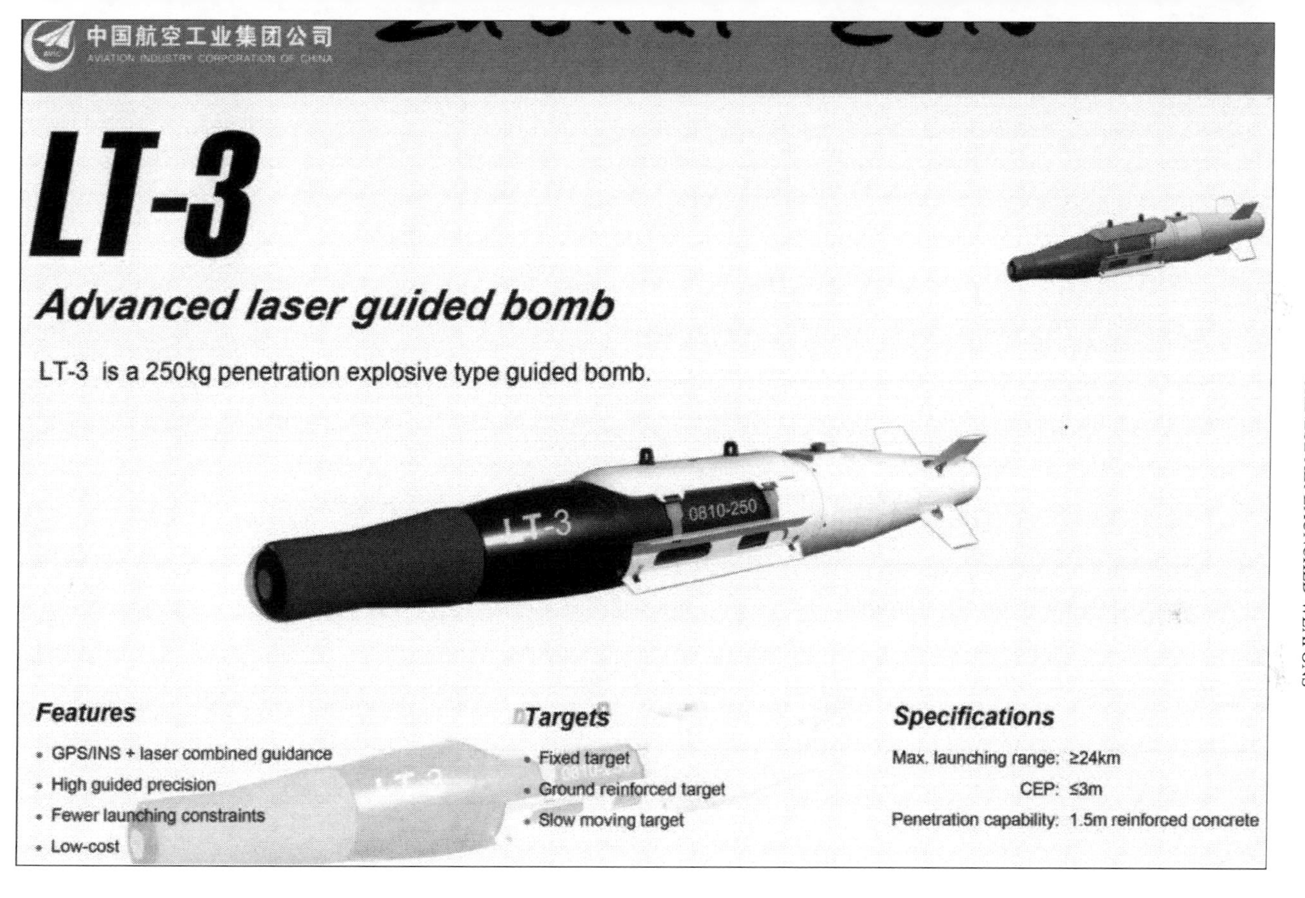

LT-3 Advanced Laser Guided Bomb, Aviation Industry Corporation of China (AVIC). 2010 Zhuhai Airshow (1/1).

General Purpose Bomb Series. Norinco Group Harbin Jiancheng Group Co. 2012 Zhuhai Airshow (1/2)

GENERAL PURPOSE
BOMB SERIES

●低阻系列 Low drag series

250kg低阻航空爆破炸弹
250kg Low Drag Aerial Bomb
长 度 Length：2100mm
重 量 Weight：216kg

500kg低阻航空爆破炸弹
500kg Low Drag Aerial Bomb
长 度 Length：2865mm
重 量 Weight：469kg

1000kg低阻航空爆破炸弹
1000kg Low Drag Aerial Bomb
长 度 Length：4000mm
重 量 Weight：932kg

●低阻低空系列 Low drag low altitude series

200kg低阻低空航空半穿甲炸弹
200kg Low Drag Low Altitude SAP Aerial Bomb
长 度 Length：2500mm
重 量 Weight：200kg

250kg低阻低空航空爆破炸弹
250kg Low Drag Low Altitude Aerial Bomb
长 度 Length：2135mm
重 量 Weight：240kg

500kg低阻低空航空爆破炸弹
500kg Low Drag Low Altitude Aerial Bomb
长 度 Length：2928mm
重 量 Weight：510kg

General Purpose Bomb Series. Norinco Group Harbin Jiancheng Group Co. 2012 Zhuhai Airshow (2/2)

Tiangang 500kg Gliding Extended Range GPS Guided Bomb. Norinco Group Harbin Jiancheng Group Co. 2012 Zhuhai Airshow (1/2).

TIANGANG

500KG GLIDING EXTENDED RANGE GPS GUIDED BOMB

系列500公斤滑翔增程卫星制导炸弹

●用途和特点

用于中高空精确攻击指挥所、通讯枢纽、军事基地、导弹基地、各类大中型桥梁、发电站、仓库等地（水）面固定点目标。

●技术参数

弹 径：377mm

弹 重：590kg ± 15kg

威 力：距爆心45m处超压为25kpa，静爆弹坑抛土量47m^3(测试值)。

精 度：CEP≤15m

射 程：≥80km

Tiangang 500kg Gliding Extended Range GPS Guided Bomb. Norinco Group Harbin Jiancheng Group Co. 2012 Zhuhai Airshow (2/2).

TIANGE

SERIES LASER GUIDED BOMBS

●USE AND CHARACTERISTICS

It is mainly used to attack fixed or low-speed (less than 10Km/h) moving point and hard targets on ground or water surface, underground/semi-underground/ground communication command centre, ammunition warehouses, oil depots, aircraft shelters, large and medium-sized bridges, traffic hubs and harbours. It can be used to attack small targets also. It can be carried by UAVs and form weapon system with training planes and light combat aircrafts.

●TECHNICAL PARAMETERS

○100kg Dual Mode Laser / GPS Guided Bomb
Diameter: 228mm
Weight: 130kg ± 10kg
Hit precision: CEP ≤ 3m
Range: ≥ 30km

○250kg Gliding Extended Range Dual Mode Laser / GPS Guided Bomb
Diameter: 299mm
Weight: 275kg ± 15kg
Effectiveness: The overpressure is 26kpa at the position 22m away from the explosion centre (by testing).
Hit precision: CEP ≤ 4m
Range: ≥ 80km

○250kg Dual Mode Laser / GPS Guided Bomb
Diameter: 299mm
Weight: 260kg ± 15kg
Effectiveness: The overpressure is 26kpa at the position 22m away from the explosion centre (by testing).
Hit precision: CEP ≤ 4m
Range: ≥ 20km

○500kg Laser Guided Bomb
Diameter: 377mm
Weight: 572kg ± 15kg
Effectiveness: The overpressure is 25kpa at the position 45m away from the explosion centre.
Hit precision
Ground laser illumination: CEP ≤ 3m
Aircraft illumination: CEP ≤ 5m

○1000kg Dual Mode Laser / GPS Guided Bomb
Diameter: 377mm/315mm/460mm
Weight: 1050kg ± 20kg
Warhead Type: HE, Penetrator, Deep penetrator
Effectiveness: It can penetrate not less than 2.4m thick reinforced concrete with 35MPa strength.
Hit precision: CEP ≤ 3m
Range: ≥ 20km

Tiange Series Laser Guided Bombs. Norinco Group Harbin Jiancheng Group Co. 2012 Zhuhai Airshow (1/2).

TIANGE
SERIES LASER GUIDED BOMBS
系列激光制导炸弹

●用途和特点

主要用于临空攻击地（海）面固定或低速运动(目标速度小于30Km/h)的点、硬目标，地下/半地下/地面通信指挥中心、武器库、油库、飞机掩体、大中型桥梁、交通枢纽、码头等重要点硬目标，兼顾攻击小目标。可用于无人机挂载，并可与教练机、轻型作战飞机构成武器系统。

●100公斤激光制导炸弹技术参数

弹 径：228mm
弹 重：130kg±10kg
精 度：CEP≤3m
射 程：≥30km

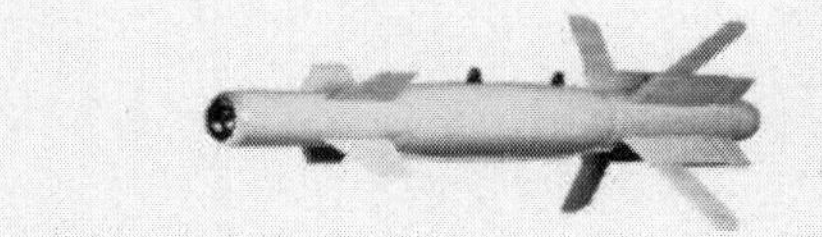

●250公斤激光制导炸弹技术参数

弹 径：299mm
弹 重：260kg±15kg
威 力：距爆心22m处超压为26kpa(测试值)。
精 度：CEP≤4m
射 程：≥20km

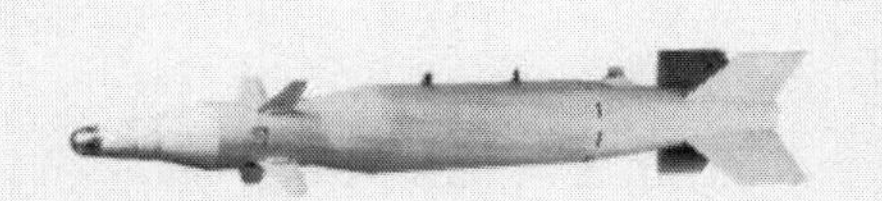

●250公斤滑翔增程激光制导炸弹技术参数

弹 径：299mm
弹 重：275kg±15kg
威 力：距爆心22m处超压为26kpa(测试值)。
精 度：CEP≤4m
射 程：≥80km

●500公斤激光制导炸弹技术参数

弹 径：377mm
弹 重：572kg±15kg
威 力：距爆心45m处超压为25kpa，
地照精度：CEP≤3m
空照精度：CEP≤5m

●1000公斤激光制导炸弹技术参数

弹 径：377mm/315mm/460mm
弹 重：1050kg±20kg
战斗部种类：爆破、侵彻、深侵彻
威 力：贯穿35兆帕钢筋混凝土厚度不小于2.4米。
精 度：CEP≤3m
射 程：≥20km

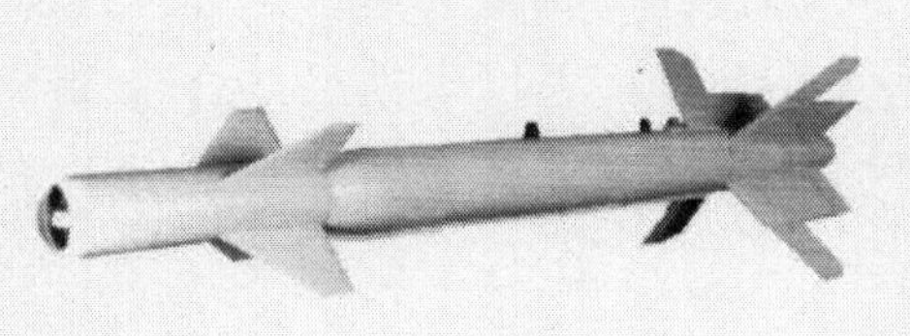

Tiange Series Laser Guided Bombs. Norinco Group Harbin Jiancheng Group Co. 2012 Zhuhai Airshow (2/2).

●**USE AND CHARACTERISTICS**

90mm airborne rocket is an air–to–ground rocket, which destroys the following targets by fragments and shockwave produced from explosion of warhead: personnel and technical weapons on ground, especially suitable to attack personnel and technical weapons in semi–open space and complicated landform, such as parking apron, oil depot, artillery position, firepower spots, facilities on ground, assembled infantry, light armored vehicles, radar, other battle field targets and small–sized ships on water surface. It can be also used to attack air targets. Guidance for the rockets can be conducted.

●**TECHNICAL PARAMETERS**

Diameter: 90mm

Weight: 16.8kg

Firing range: 8000m

TianJian Series 90mm Airborne Rocket. Norinco Group Harbin Jiancheng Group Co. 2012 Zhuhai Airshow (1/2).

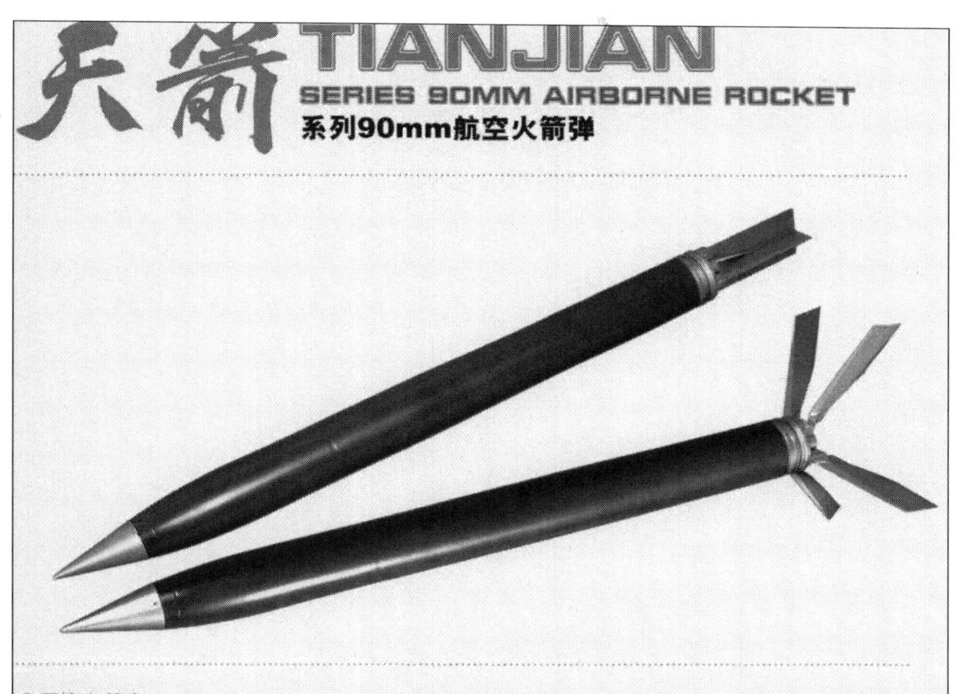

●用途和特点

90mm航空火箭弹是一种空地火箭弹，利用其战斗部爆炸时产生的破片和冲击波来毁伤目标，打击地面的人员和各种技术兵器，特别适用于打击半敞开空间或复杂地形内的人员和技术兵器。如：机场停机坪、油库、火力点、地面设施、步兵群、轻型装甲车辆、雷达、炮兵阵地及其它各种战场目标和小型舰艇等水上目标，亦可打击空中目标。可以进行制导化改造。

●技术参数

弹 径：90mm
弹 重：16.8kg
射 程：8000m

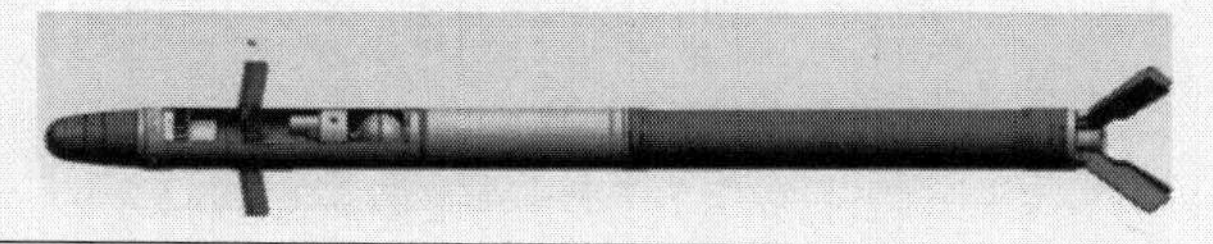

TianJian Series 90mm Airborne Rocket. Norinco Group Harbin Jiancheng Group Co. 2012 Zhuhai Airshow (2/2).

●用途和特点

布撒武器是一种由作战飞机挂载、远程投放、隐身突防、全天候使用、多用途、滑翔型、精确到达目标上空布撒多种子弹药的空地制导武器，主要用于攻击敌机场跑道、技术兵器阵地、停机坪上的飞机、电力设施、集群武装人员等各类面目标。

●技术参数

外型尺寸：≤4500mm×600mm×470mm

弹 重：≤680kg

精 度：子弹药散布中心精度CEP≤35m

（条件：相对目标高度200m以下，最大风速≤10m/s）

Tianlei Series 500kg Airborne Dispenser Weapon. Norinco Group Harbin Jiancheng Group Co. 2012 Zhuhai Airshow (1/2).

TIANLEI

SERIES 500KG AIRBORNE DISPENSER WEAPON

系列500公斤航空布撒武器

●USE AND CHARACTERISTICS

The airborne dispenser weapon is an air–to–ground guided one, which is carried by combat aircrafts and released from long distance. This weapon is characterized by defense penetration, all–weather employment, multipurpose, glide type, approaching precisely to the above of the targets and dispensing multipurpose bomblets. This weapon is mainly used to attack surface distributive targets like airport runway, technical weapon position, aircrafts at landing field, electric power facilities and assembled armed personnel.

●TECHNICAL PARAMETERS

OVERALL DIMENSIONS:

≤4500mm × 600mm × 470mm

WEIGHT: ≤680kg

HIT PRECISION:

The hit precision of bomblets dispersion center is CEP≤35m (with altitude relative to target of less than 200m and max. wind speed of equal to or less than 10m/s)

Tianlei Series 500kg Airborne Dispenser Weapon. Norinco Group Harbin Jiancheng Group Co. 2012 Zhuhai Airshow (2/2).

Bomb REK

Brief Introduction for NAV-SDB Small Diameter Bomb

NAV-SDB small diameter gliding bomb is a new type of miniaturized standoff range guided munition developed from the US made model of SDB. With the characteristics of small size and light weight, SDB is suitable for being mounted on a variety of aircraft (manned or unmanned) to improve single bomb load and reduce flight sorties; SDB with extended gliding assembly to extend standoff range can greatly improve the aircraft safety; with combined inertial and CNSS guidance, it has high hit accuracy; and with the characteristics of low explosive charge and thus less collateral damage, it has become the ideal weapon to be used in urban combat.

Application & Mission Modes

- Application
  NAV-SDB small diameter gliding bombs are mainly carried by combat aircrafts for fighting the enemy's light armored vehicles, air defense missile position, technical vehicles, planes on the tarmac, air defense radar, important targets in the city, and effective strength etc. And during combat the bombs are usually released outside of the intermediate and short range ground air defense fire area.

- Mission Modes
  There are two modes of combat: coordinate attack and direct aiming attack; coordinate attack is the main mode. The bombs will fly with autonomous guidance after launching.

Operational Performance and Tactic & Technical Specification

- Suitable Aircrafts
  Designed with standardized interface, NAV-SDB small diameter gliding bomb is suitable for A-5, J-7, JF-17, JH-7, J-10, J-11B, and J-20 etc. manned and high speed unmanned aircrafts with weapon systems in accordance with the standard requirements.

- Physical specification
  Length : Less than 2 m ;
  Diameter : 185 mm ;
  Weight : Less than 120 kg ;
  Explosive charge : more than 35 kg ;
  Wing span : 1800 mm.

NAV-SDB Small Diameter Bomb (Bomb REK). NAV Technology Co. 2014 Zhuhai Airshow (1/2).

Bomb REK

- Realese condition
  Realese altitude：3000 m—12000 m
  Realese speed：0.6 –1.5 Mach

- Flying Range

Realese Altitude	12000 m	4000 m
Realese Velocity	900 km/h	900 km/h
Flying range, no less than (In horizontal Realese)	80 km	20 km

- Accuracy
  Accuracy：13 m (CEP)

NAV-DB Small Diameter Bomb (Bomb REK). NAV Technology Co. 2014 Zhuhai Airshow (2/2).

Brief Introduction for
NAV-REK Range-Extended guided Bomb

General

Precision strike has become the main type of modern local war. Since the last 20 years, precision-guided bombs have been widely applied and made significant gains in Gulf war, Kosovo war, Afghanistan war, Iraq war and the recent Libya war. NAV-REK Range-Extended guided Bomb adopts "inertial navigation+ GNSS correction" guidance system, is an all-weather operational capability guided bomb. Compared with ordinary bomb, it installs gliding extended range components, which has notable features of low cost, stand-off and fire-and-forget, has become one of the main development direction of information-based guided bomb.

NAV-REK Range-Extended guided Bomb adopts compound guidance system of strap-down inertial navigation+ GPS satellite correction. CEP is less than 15m. NAV-REK Range-Extended guided Bomb adopts high-aspect-ratio wings and lift-to-drag ratio aerodynamic design combined with "X" type tail aerodynamic form to realize that the bomb maximum firing range is wider than 80km. Warhead adopts generalization and modularization design principle. It can reload semi-armor-piercing warhead and blast warhead to meet different requirements, precision strikes on the ground target. It has advanced overall performance, similar to American extended range JDAM（JDAM-ER）.

System Features

- Adapt to various carrier
  NAV-REK Range-Extended guided Bomb has simple operation process, broaden the bomb package net, relief pilot's load to help complete task. It requires simple on carrier mount, adapts to various existing aircrafts, greatly enhance the capacity of long-range precision attack.
- Flexible in Military Operation
  NAV-REK Range-Extended guided Bomb fully adopts integrated navigation system information close-loop guidance. Compared with laser-guided bomb, it has fire-and-forget flexibility, reduces the casualty rate of ground person, improves the viability of the carrier.
- Far Attack Range
  NAV-REK Range-Extended guided Bomb adopts gliding extended range technology. Its maximum firing range is not less than 80km. So it is able to drop a bomb outside enemy's defense field, improve the carrier safety.
- Advanced Guidance Control Technology
  NAV-REK Range-Extended guided Bomb adopts "Gliding extended range components+ bomb body+ 'X' type full motion stern rudder" tailless pneumatic layout scheme. This scheme provide greater lift-to-drag ratio, while guarantee bombs have good maneuverability and stability.

NAV-REK Range-Extended Guided Bomb. NAV Technology Co. 2014 Zhuhai Airshow (1/2).

- All-weather Use

  NAV-REK Range-Extended guided Bomb fully adopts integrated navigation system information close-loop guidance. Compared with laser-guided bomb which is easy affected by weather, it has all-weather use feature, not conditioned by weather, dust and other bad factors in battlefield.

Tactical and Technical Parameters

- Attack Target

  NAV-REK Range-Extended guided Bomb can be used to attack command post, communication hinge, military base, missile base, all kinds of large and medium-sized bridges, power station, warehouse and other targets. It mainly used for attack ground (water) surface fixed target in medium-altitude.

- Attack Mode

  Coordinate attack

- Carrier

  A-5 series, JF-17, J10, K8 trainer aircraft and so on, it also can be loaded on battle aircraft to meet clients' requirement.

- Operating Conditions

  Release altitude（3000）12000m;
  Release speed（600）1000km/h

- Maximum Gliding Distance

  No less than 80km

- Guidance System and Precision

  Guidance system: "Inertial navigation+ satellite correction" combined guidance;
  Accuracy：CEP no larger than 15m。

- Main Physical Parameters

Bomb weight：	580kg；
Bomb length：	Less than 3400mm；
Maximum diameter of bomb body：	Less than 0.4m；
Wingspan：	3150 mm

- Power

  Blast warhead：TNT equivalent no less than 280kg；

- Interface

  Mechanical interface: Meet GLB1C requirement, ejection release, adopt GJB637 II lifting lug, space between lifting lugs: 355.6mm.

  Electrical Interface: Meet GJB1188A(same as MIL-1760C) requirement, adopt J599/20WJ20SN type electric connector.

- Environment

  Storage Temperature：-55℃+60℃
  Operating Temperature：-40℃+60℃

NAV-REK Range-Extended Guided Bomb. NAV Technology Co. 2014 Zhuhai Airshow (2/2).

Brief Introduction for NAV-LGB Series Laser Guided Bomb

General

With laser and GPS+INS dual guide mode, NAV-LGB Series laser guided bomb is mainly used to attack fixed or low-speed (less than 10Km/h) moving point and hard targets on ground or water surface, underground / semi-underground / ground communication command centre, ammunition warehouses, oil depots, aircraft shelters, large and medium-sized bridges, traffic hubs and harbours. It can be used to attack small targets also. It can be carried by UAVs and form weapon system with training planes and light combat aircrafts.

Tactical and Technical Parameters

- Parameters

PARAMETERS	250Kg	500Kg
Release Height	3000m - 12,000m	
precision	≤ 4 m(CEP)	Ground Illumination ≤ 3 m(CEP) Aircraft Illumination ≤ 5 m(CEP)
Range	20km	20km
Diameter	299 mm	377 mm
Weight	Less than 275 kg	Less than 275 kg
Effectiveness	The overpressure is 26KPa at the position 22m away from the explosion center(By test)	The overpressure is 25KPa at the position 45m away from the explosion center(By test)

- Operating Conditions
  Release altitude：3000～12000m;
  Release speed： 600～1000km/h。
- Interface
  Mechanical interface: Meet GLB1C requirement, ejection release, adopt GJB637 II lifting lug, space between lifting lugs: 355.6mm.
  Electrical Interface: Meet GJB1188A(same as MIL-1760C) requirement, adopt J599/20WJ20SN type electric connector.
- Environment
  Storage Temperature：-55℃+60℃
  Operating Temperature：-40℃+60℃

NAV-LGB Series Laser Guided Bomb. NAV Technology Co. 2014 Zhuhai Airshow (1/2).

- All-weather Use

  NAV-REK Range-Extended guided Bomb fully adopts integrated navigation system information close-loop guidance. Compared with laser-guided bomb which is easy affected by weather, it has all-weather use feature, not conditioned by weather, dust and other bad factors in battlefield.

Tactical and Technical Parameters

- Attack Target
  NAV-REK Range-Extended guided Bomb can be used to attack command post, communication hinge, military base, missile base, all kinds of large and medium-sized bridges, power station, warehouse and other targets. It mainly used for attack ground (water) surface fixed target in medium-altitude.
- Attack Mode
  Coordinate attack
- Carrier
  A-5 series, JF-17, J10, K8 trainer aircraft and so on, it also can be loaded on battle aircraft to meet clients' requirement.
- Operating Conditions
  Release altitude（3000）12000m;
  Release speed（600）1000km/h
- Maximum Gliding Distance
  No less than 80km
- Guidance System and Precision
  Guidance system: "Inertial navigation+ satellite correction" combined guidance;
  Accuracy：CEP no larger than 15m。
- Main Physical Parameters

Bomb weight：	580kg；
Bomb length：	Less than 3400mm；
Maximum diameter of bomb body：	Less than 0.4m；
Wingspan：	3150 mm

- Power
  Blast warhead：TNT equivalent no less than 280kg；
- Interface
  Mechanical interface: Meet GLB1C requirement, ejection release, adopt GJB637 II lifting lug, space between lifting lugs: 355.6mm.
  Electrical Interface: Meet GJB1188A(same as MIL-1760C) requirement, adopt J599/20WJ20SN type electric connector.
- Environment
  Storage Temperature：-55℃+60℃
  Operating Temperature：-40℃+60℃

NAV-LGB Series Laser Guided Bomb. NAV Technology Co. 2014 Zhuhai Airshow (2/2).

PODS

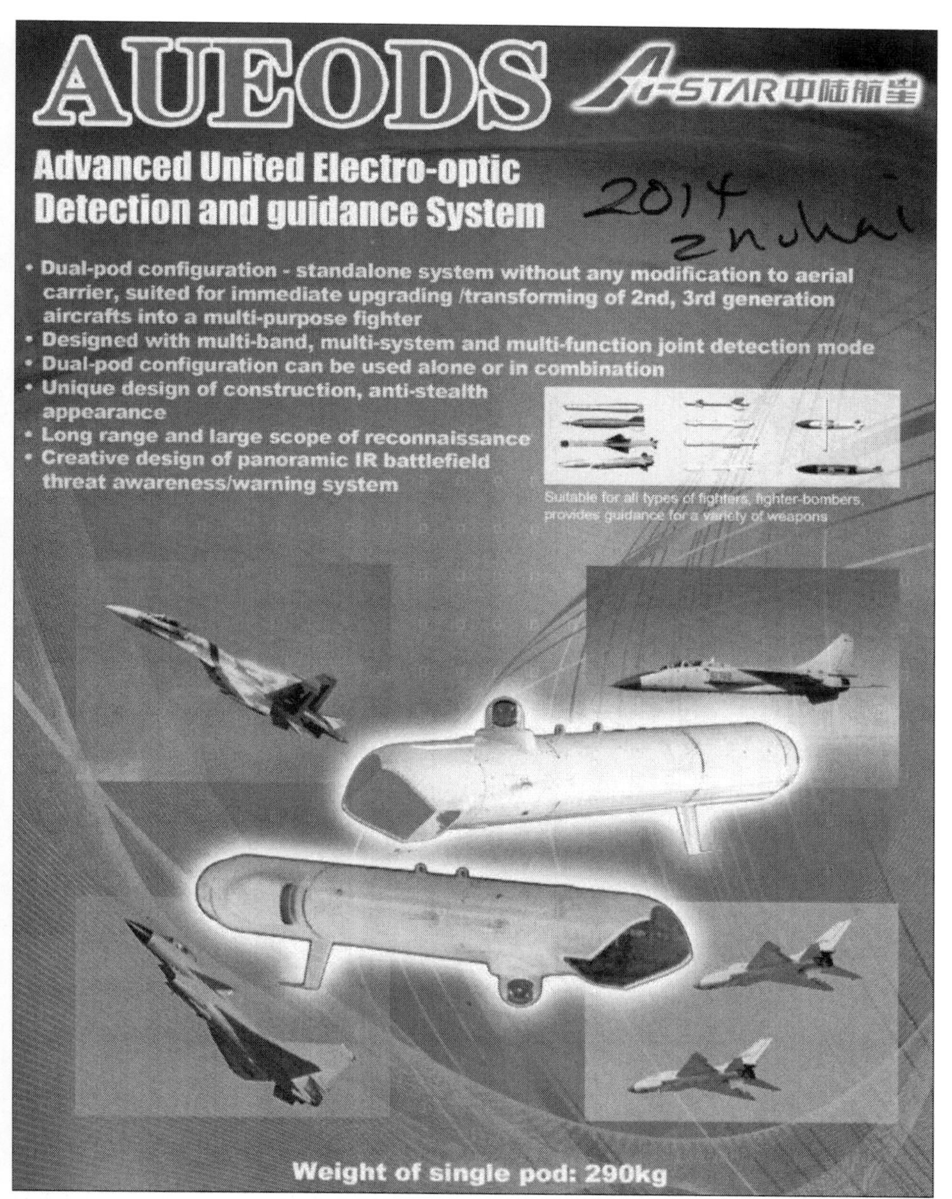

AUEODS (A-Star) Advanced United Electro-Optic Detection and Guidance System. 2014 Zhuhai Airshow (1/8).

Brief

AUEODS designed with dual-pod configuration integrates long-wave/medium-wave infrared system, laser illuminator/ranger system, up/down-looking panoramic IR awareness system, capable of performing diversified functions, such as multiband, multisystem united electro-optic detection, target tracking/positioning, threats awareness/warning, navigation/guidance, etc.

Section No.2
- warning against incoming missile
- long-range panoramic IR searching, battlefield situation awareness

Section No.4
- environment control
- two-way data link

Section No.3
- status management
- integrated information processing
- whole cabin self-test
- image/data record
- incoming threat assessment
- fire control calculation, aircraft interface
- sensors as INS, air computer, altimeter, etc.
- power unit, data transmission unit

Section No.1
- missile/bomber guidance
- searching/tracking/reconnaissance
- target position indication

Features and functions

(1) Dual-pod configuration - standalone system without any modification to aerial carrier, suited for immediate upgrading /transforming of 2nd, 3rd generation aircrafts into a multi-purpose fighter

(2) Designed with multi-band, multi-system and multi-function joint detection mode

(3) Dual-pod configuration can be used alone or in combination

(4) Unique design of construction, anti-stealth appearance

Head of AUEODS pods

Wedge-shaped head of Sniper Targeting Pod, USA

AUEODS (A-Star) Advanced United Electro-Optic Detection and Guidance System. 2014 Zhuhai Airshow (2/8).

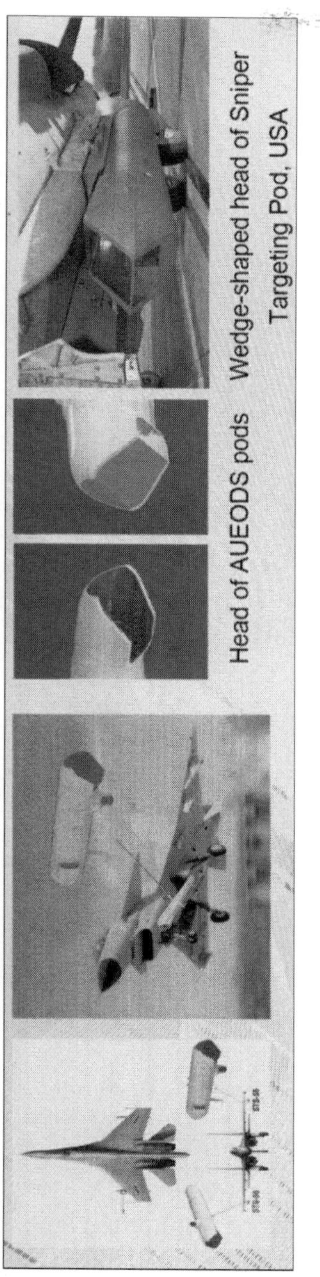

AUEODS (A-Star) Advanced United Electro-Optic Detection and Guidance System. 2014 Zhuhai Airshow (3/8). CLOSE-UP.

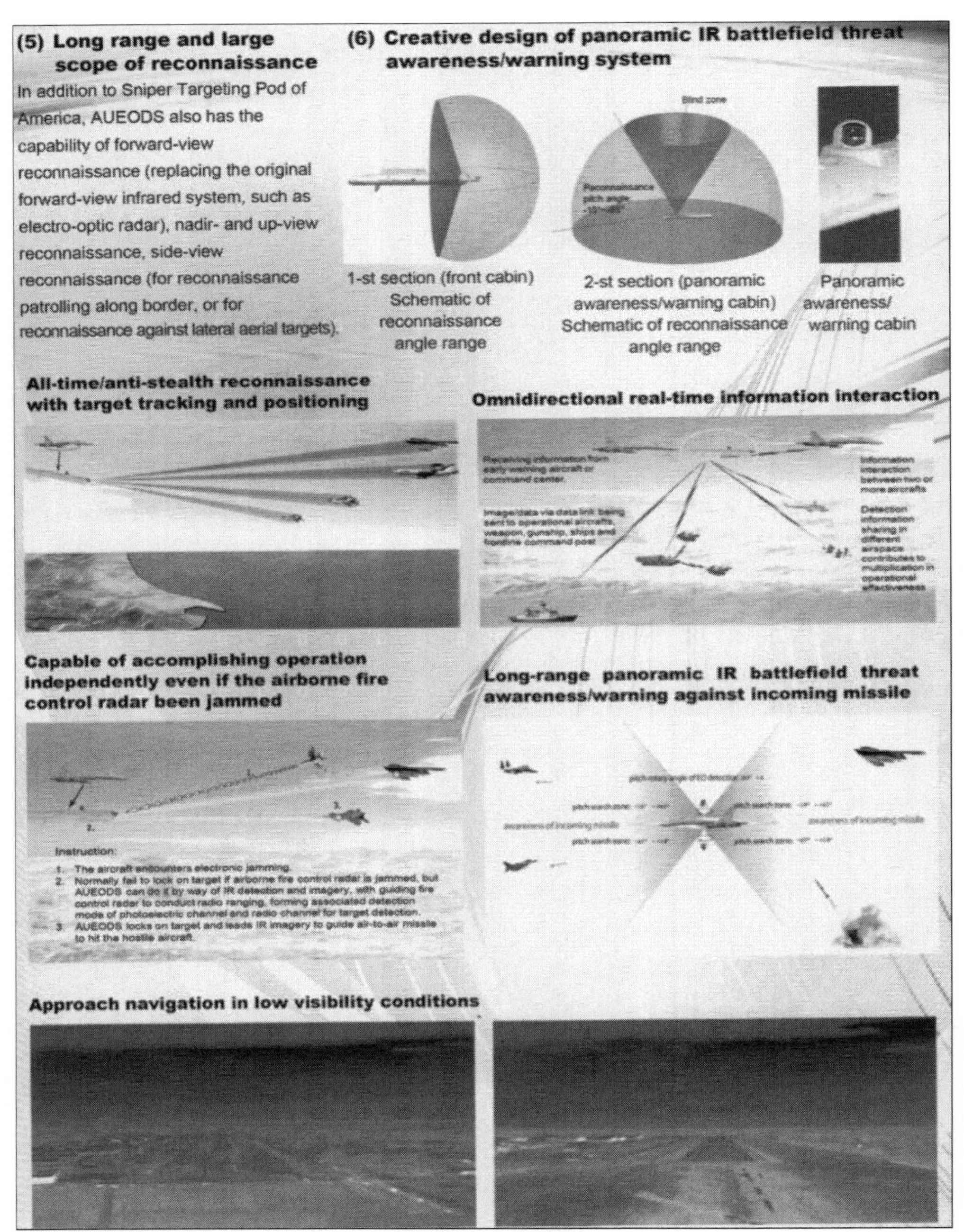

AUEODS (A-Star) Advanced United Electro-Optic Detection and Guidance System. 2014 Zhuhai Airshow (4/8). CLOSE-UP.

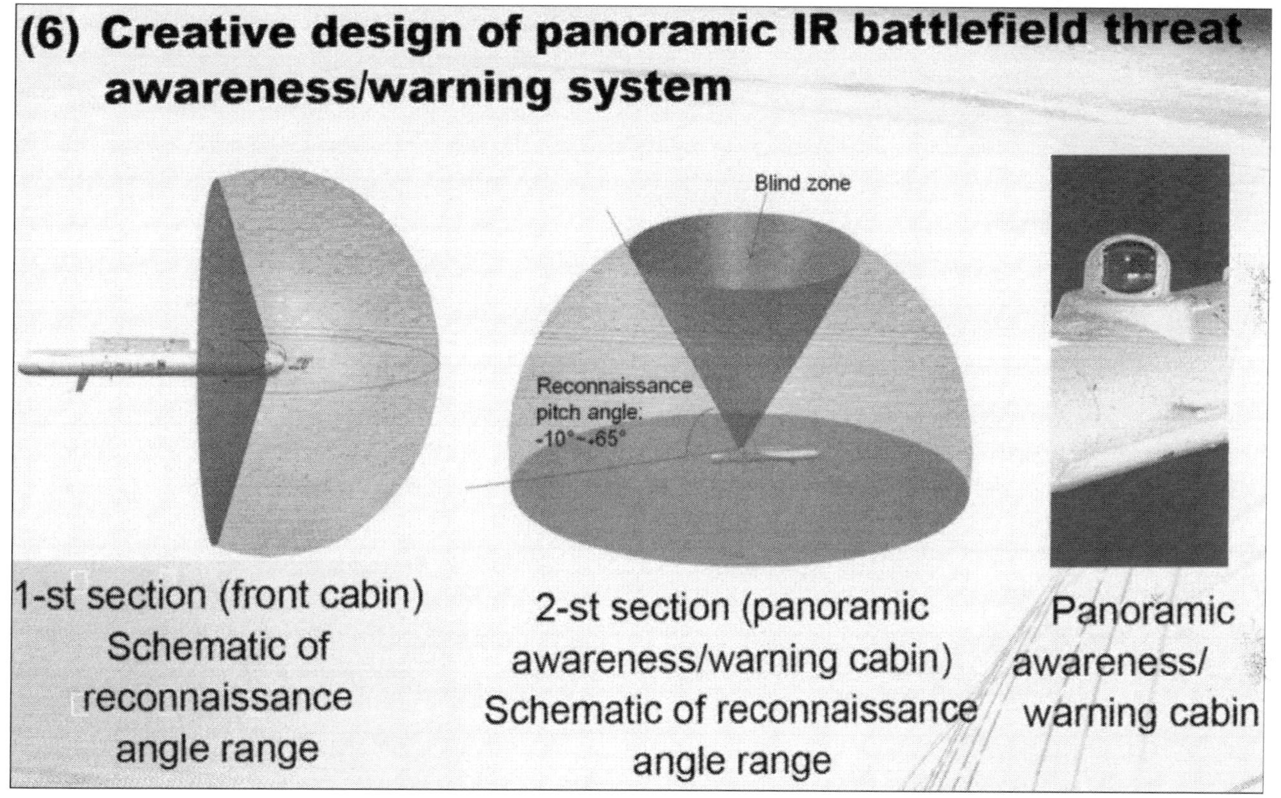

AUEODS (A-Star) Advanced United Electro-Optic Detection and Guidance System. 2014 Zhuhai Airshow (5/8). CLOSE-UP.

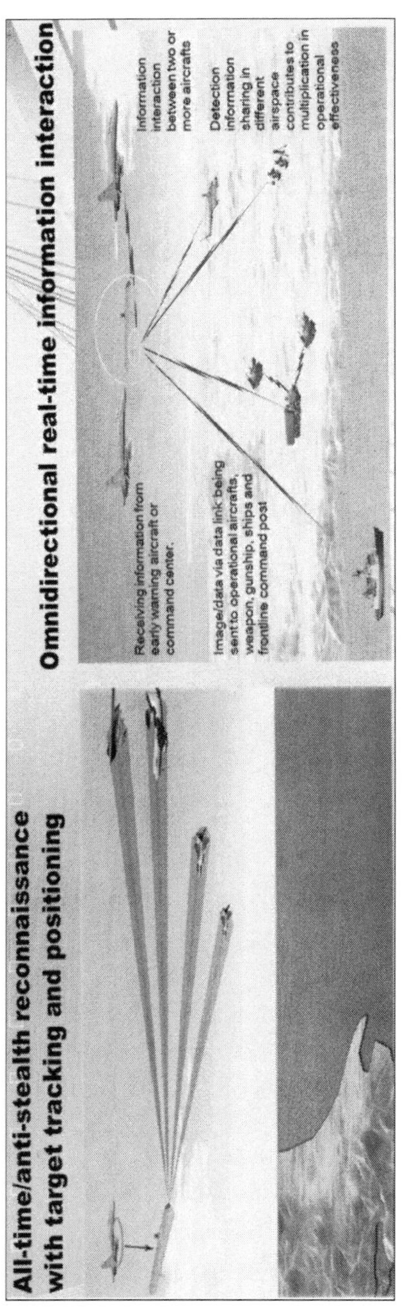

AUEODS (A-Star) Advanced United Electro-Optic Detection and Guidance System. 2014 Zhuhai Airshow (6/8). CLOSE-UP.

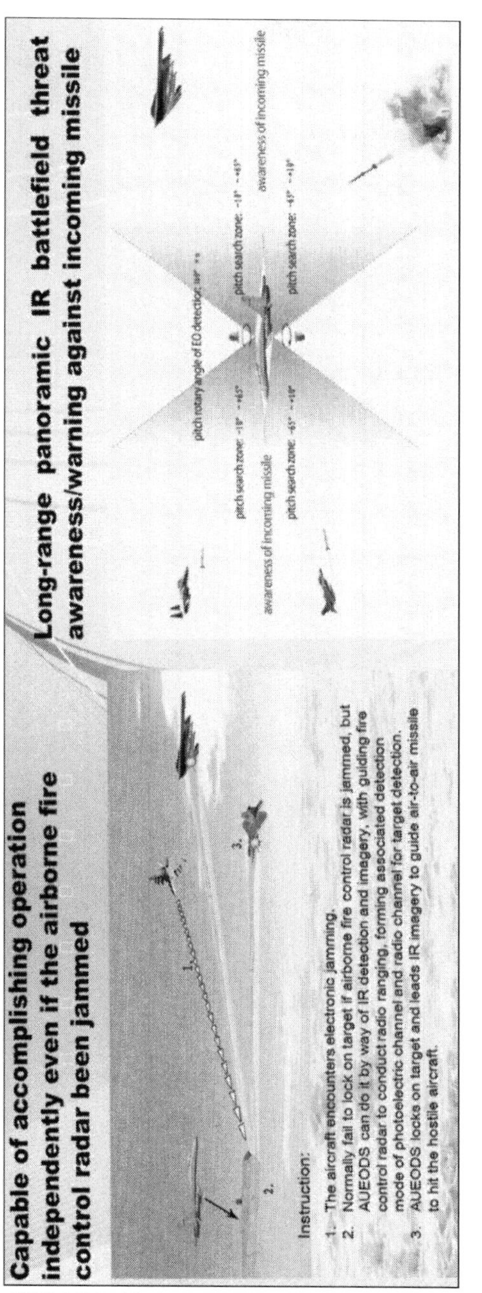

AUEODS (A-Star) Advanced United Electro-Optic Detection and Guidance System. 2014 Zhuhai Airshow (7/8). CLOSE-UP.

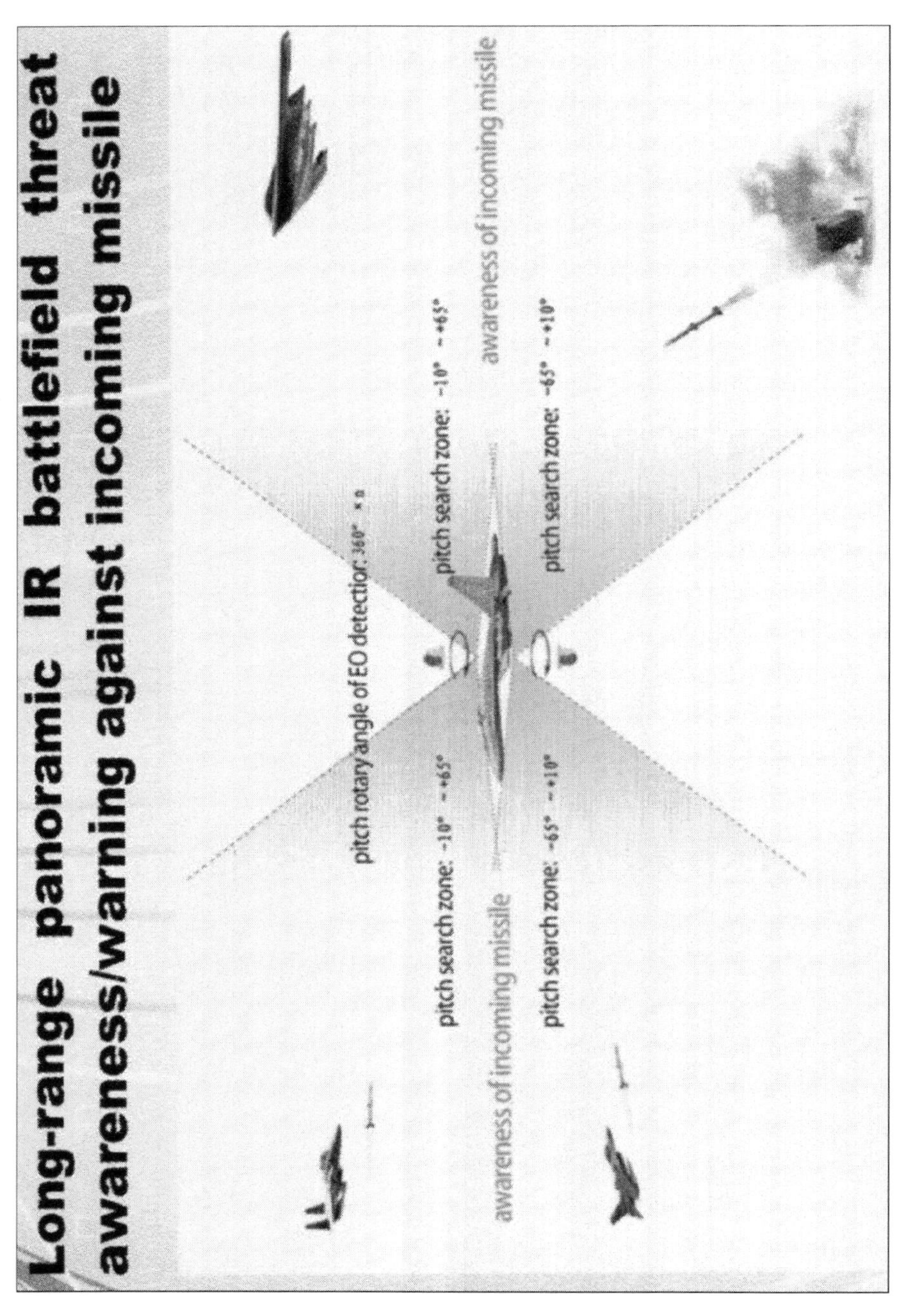

AUEODS (A-Star) Advanced United Electro-Optic Detection and Guidance System. 2014 Zhuhai Airshow (8/8). CLOSE-UP.

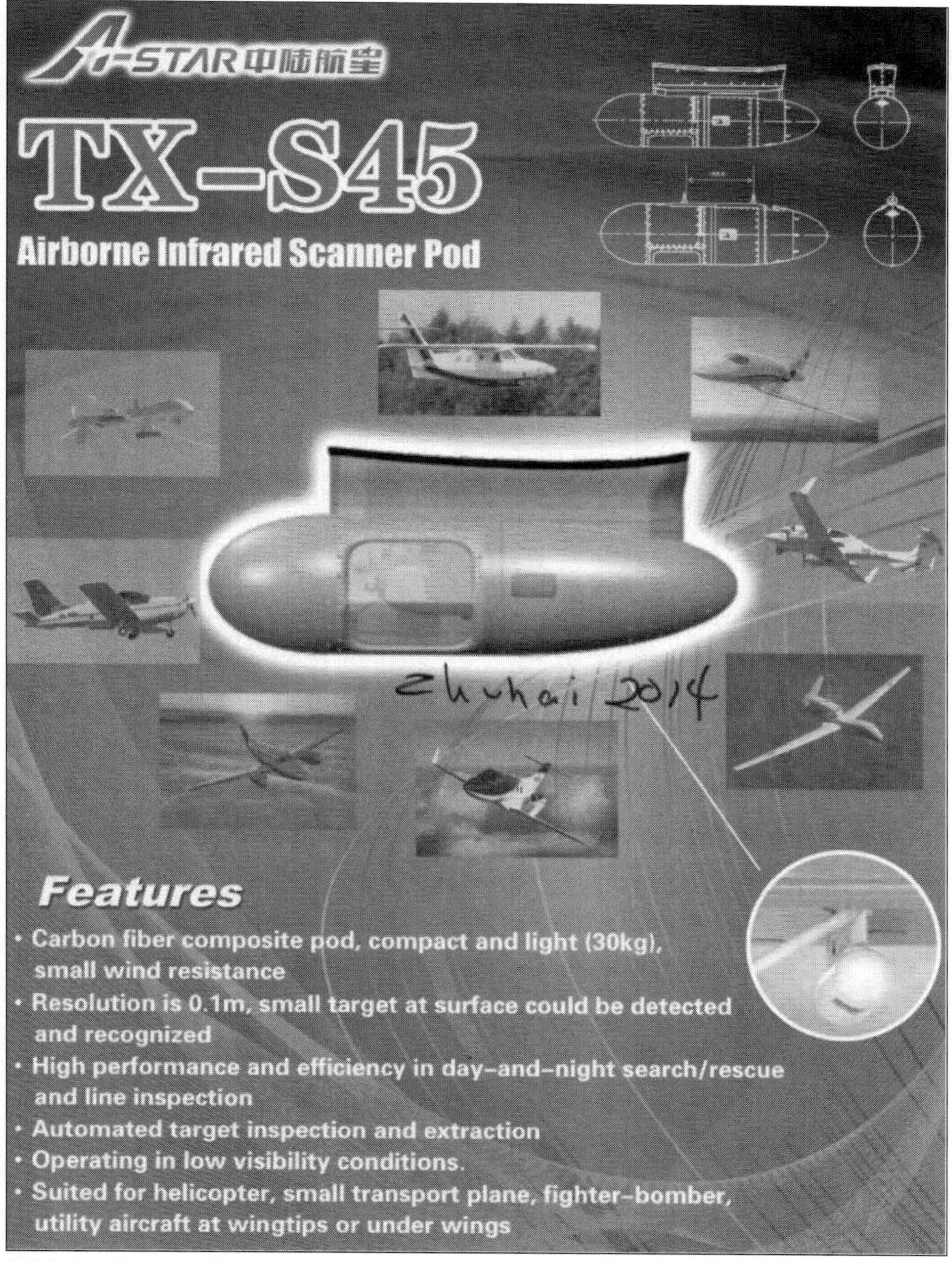

TX-S45. (A-Star) Airborne Infrared Scanner Pod. 2014 Zhuhai Airshow (1/7).

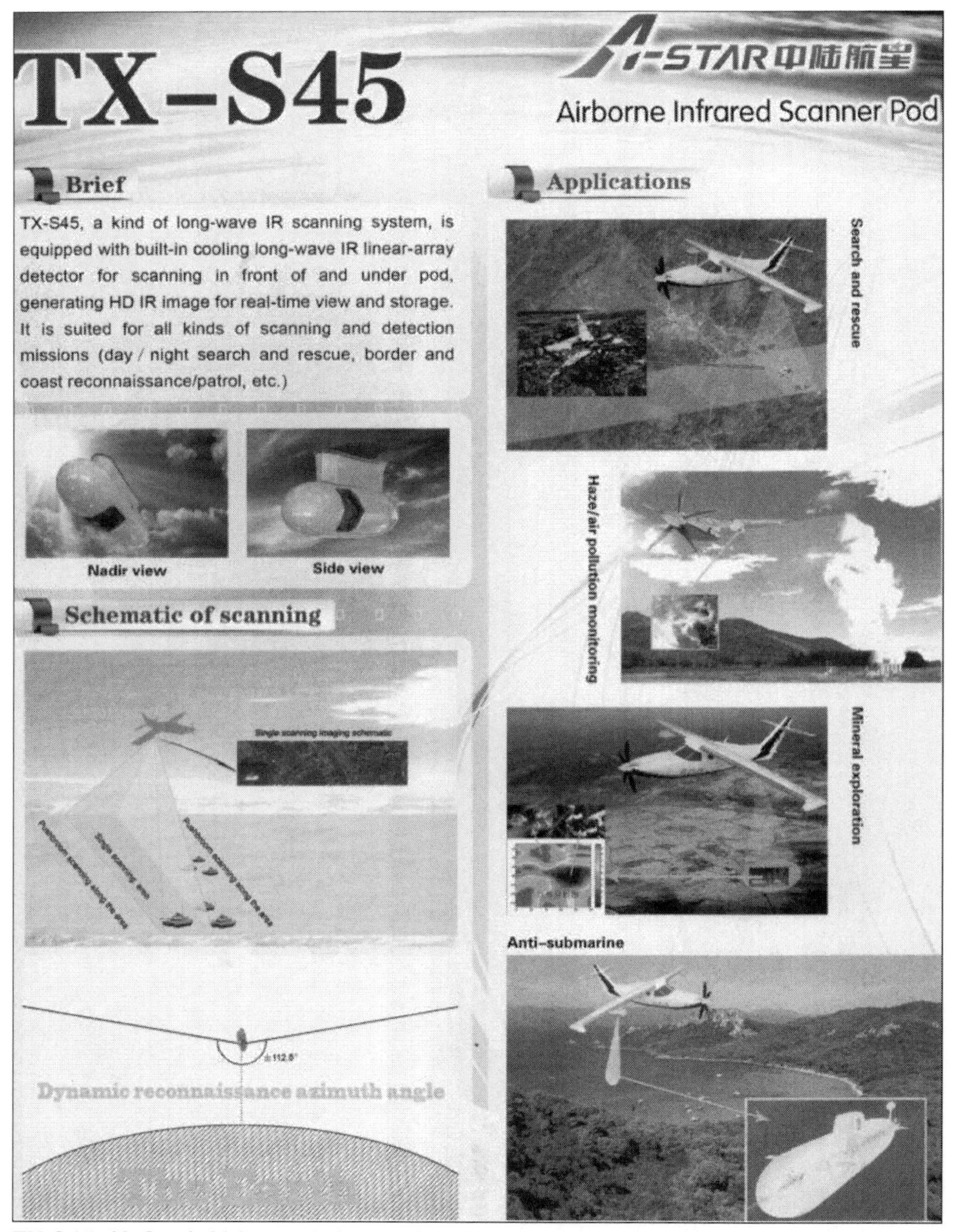

TX-S45. (A-Star) Airborne Infrared Scanner Pod. 2014 Zhuhai Airshow (2/7). CLOSE-UP.

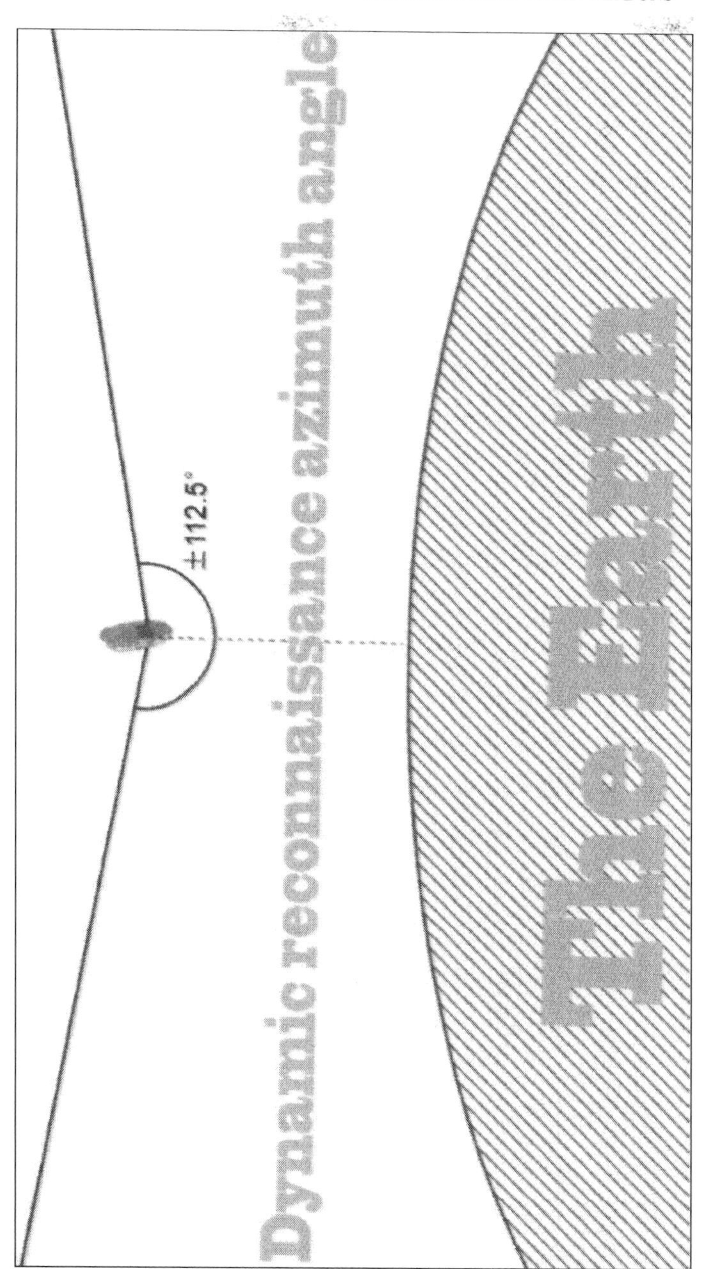

TX-S45. (A-Star) Airborne Infrared Scanner Pod. 2014 Zhuhai Airshow (3/7). CLOSE-UP.

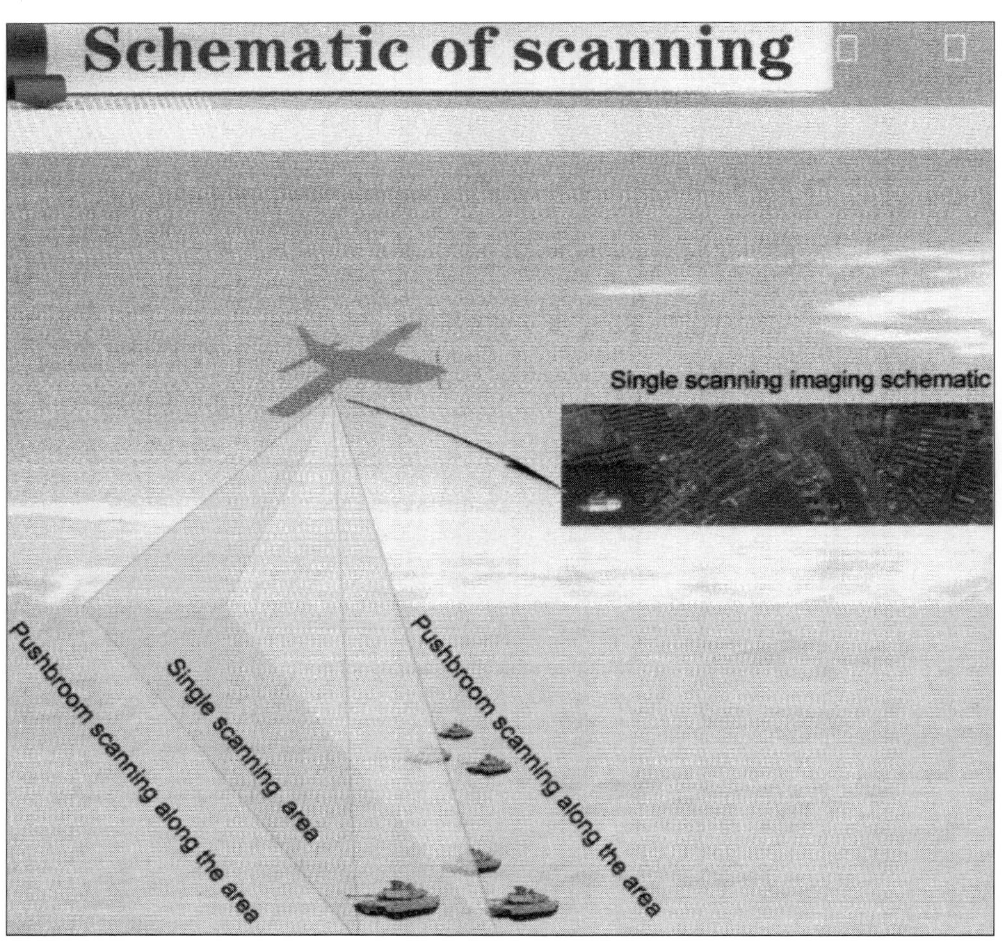

TX-S45. (A-Star) Airborne Infrared Scanner Pod. 2014 Zhuhai Airshow (4/7). CLOSE-UP.

TX-S45. (A-Star) Airborne Infrared Scanner Pod. 2014 Zhuhai Airshow (5/7). CLOSE-UP.

TX-S45. (A-Star) Airborne Infrared Scanner Pod. 2014 Zhuhai Airshow (6/7). CLOSE-UP.

TX-S45. (A-Star) Airborne Infrared Scanner Pod. 2014 Zhuhai Airshow (7/7). CLOSE-UP.

TX-S46. (A-Star) Airborne Infrared Reconnaissance Pod. 2014 Zhuhai Airshow (1/2).

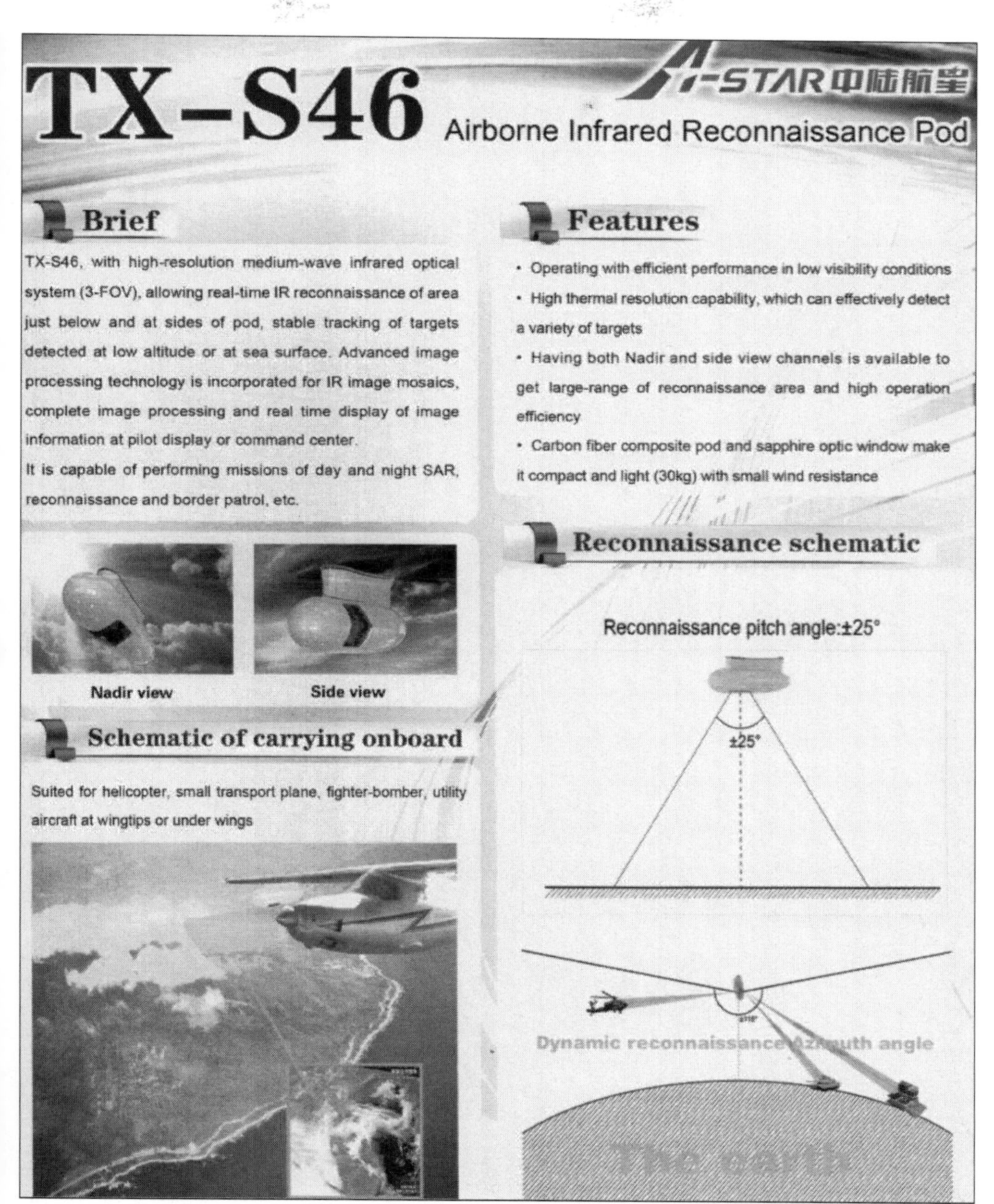

TX-S46. (A-Star) Airborne Infrared Reconnaissance Pod. 2014 Zhuhai Airshow (2/2).

TX-S50. (A-Star) Airborne Forward-Looking Infrared Scanner Pod. 2014 Zhuhai Airshow (1/4).

TX-S50. (A-Star) Airborne Forward-Looking Infrared Scanner Pod. 2014 Zhuhai Airshow (2/4).

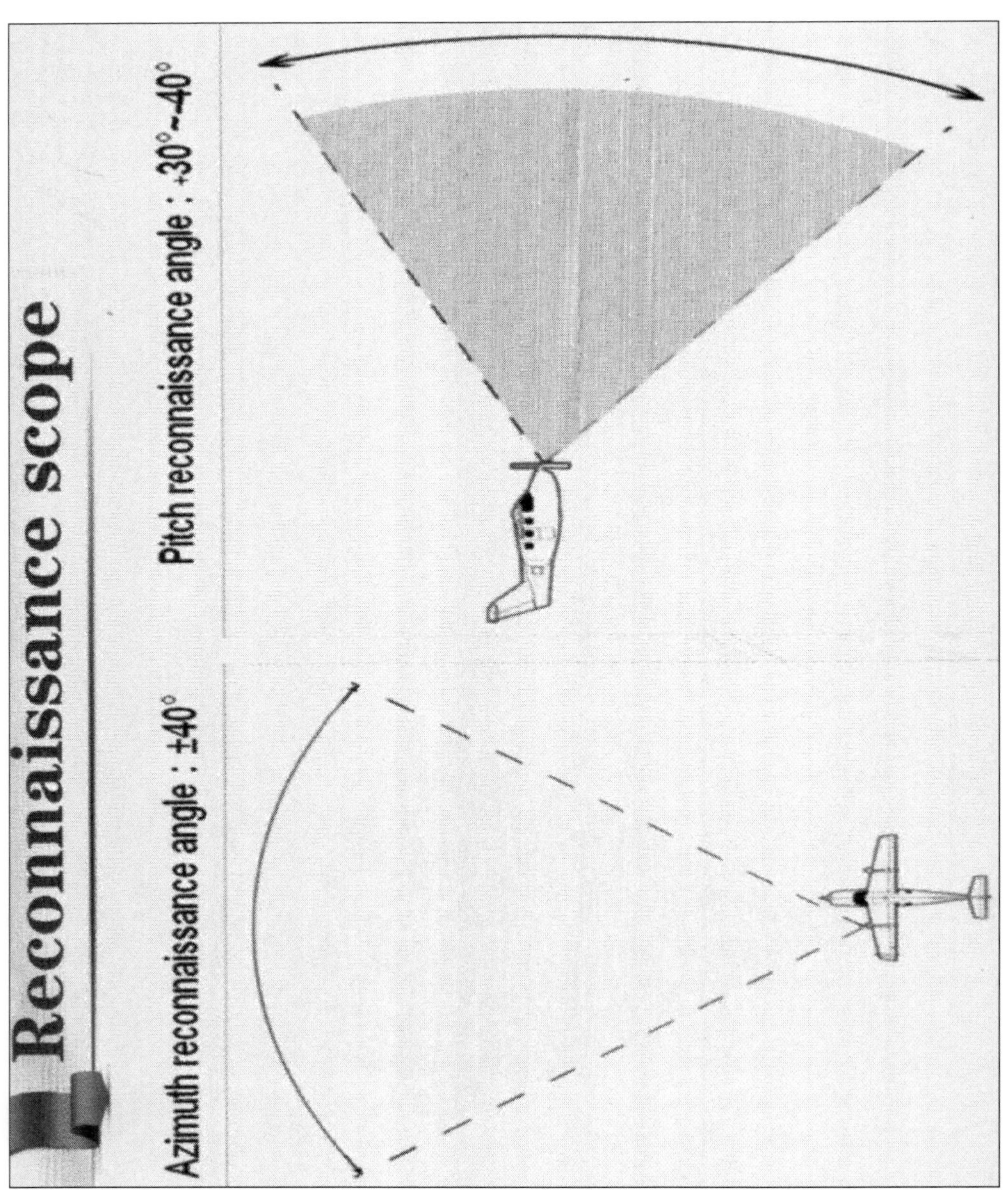

TX-S50. (A-Star) Airborne Forward-Looking Infrared Scanner Pod. 2014 Zhuhai Airshow (3/4).

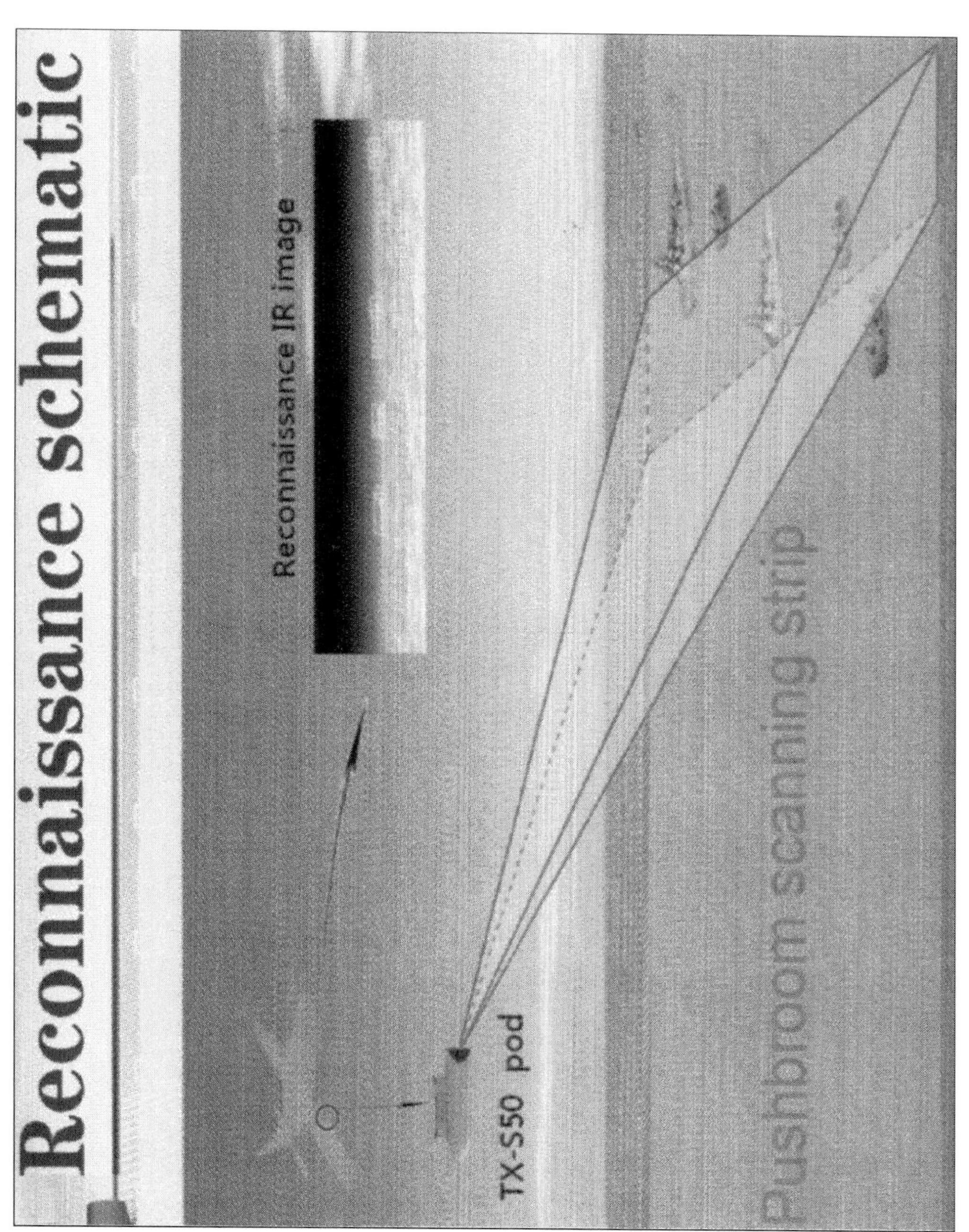

TX-S50. (A-Star) Airborne Forward-Looking Infrared Scanner Pod. 2014 Zhuhai Airshow (4/4).

TX-S63. (A-Star) Airborne 3D Digital Aerial Camera (Pod-Type). 2014 Zhuhai Airshow (1/2).

TX-S63

Airborne 3D digital aerial camera (pod-type)

Brief

Airborne 3D digital aerial camera (pod-type) represents a creative design with pushbroom airborne digital scanner imbedded in a pod, which provides easy and quick installation without any major modification to carrier aircraft, as usually conventional aerial camera does (open a hole in the belly).

Built-in 3D aerial camera is a pushbroom airborne scanner that provides high resolution imaging by three simultaneous color channels. Nadir channel captures ground surface images just below the aircraft and is used for automated creation of orthophoto. Two other channels (forward and backward) capture images with 16° and 26° angles along the flight direction providing permanent triple overlap for stereo mapping, DTM generation, etc.

Can be used for mapping, weapons guidance etc.

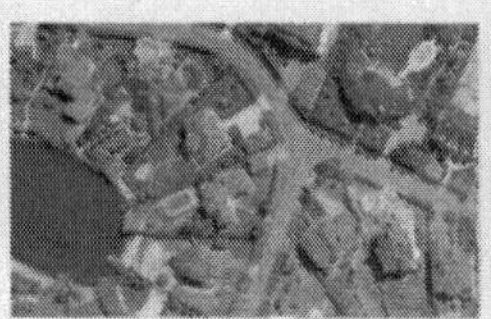

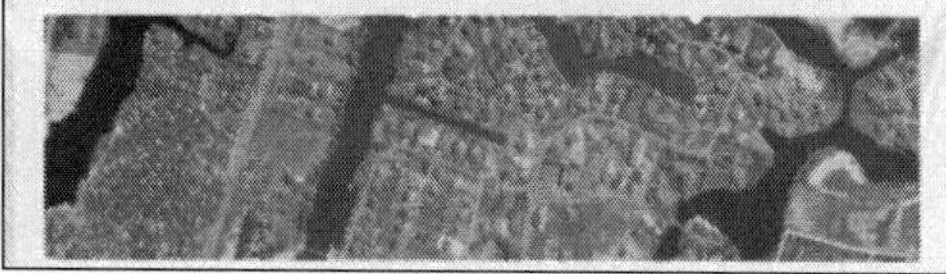

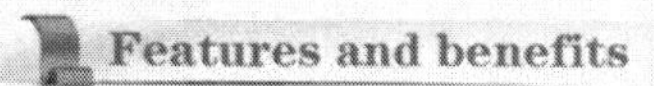

Features and benefits

Pod

- Creative design of pod-type 3D aerial camera make the concept become true
- Suited for mounting on helicopter, small transport plane, fighter-bomber, utility aircraft at wingtips or under wings without any destructive modification to carrier aircraft and any affects on the structure of the carrier aircraft, the flight envelope and cabin pressurization
- Pod shell is made of carbon fiber composite, which is compact and light (≤55kg) with less wind resistance and advanced performance
- Easy and simple operation and maintenance
- High-efficiency mapping with low operation cost

Built-in aerial camera

- Complete digital photogrammetric workflow without film development, scanning, etc.
- Three RGB-sensors provide crystal bright 42-bit images
- Contiguous seamless images for whole strip with permanent triple overlay
- Real-time image view and automatic selection of optimal exposition during the flight
- On-the-fly lossless compression allows up to 16 hours continuous capture
- Selectable stereo for 3D mapping with 16° 26° or 45° convergence angle
- Simple and robust design for easy maintenance

TX-S63. (A-Star) Airborne 3D Digital Aerial Camera (Pod-Type). 2014 Zhuhai Airshow (2/2).

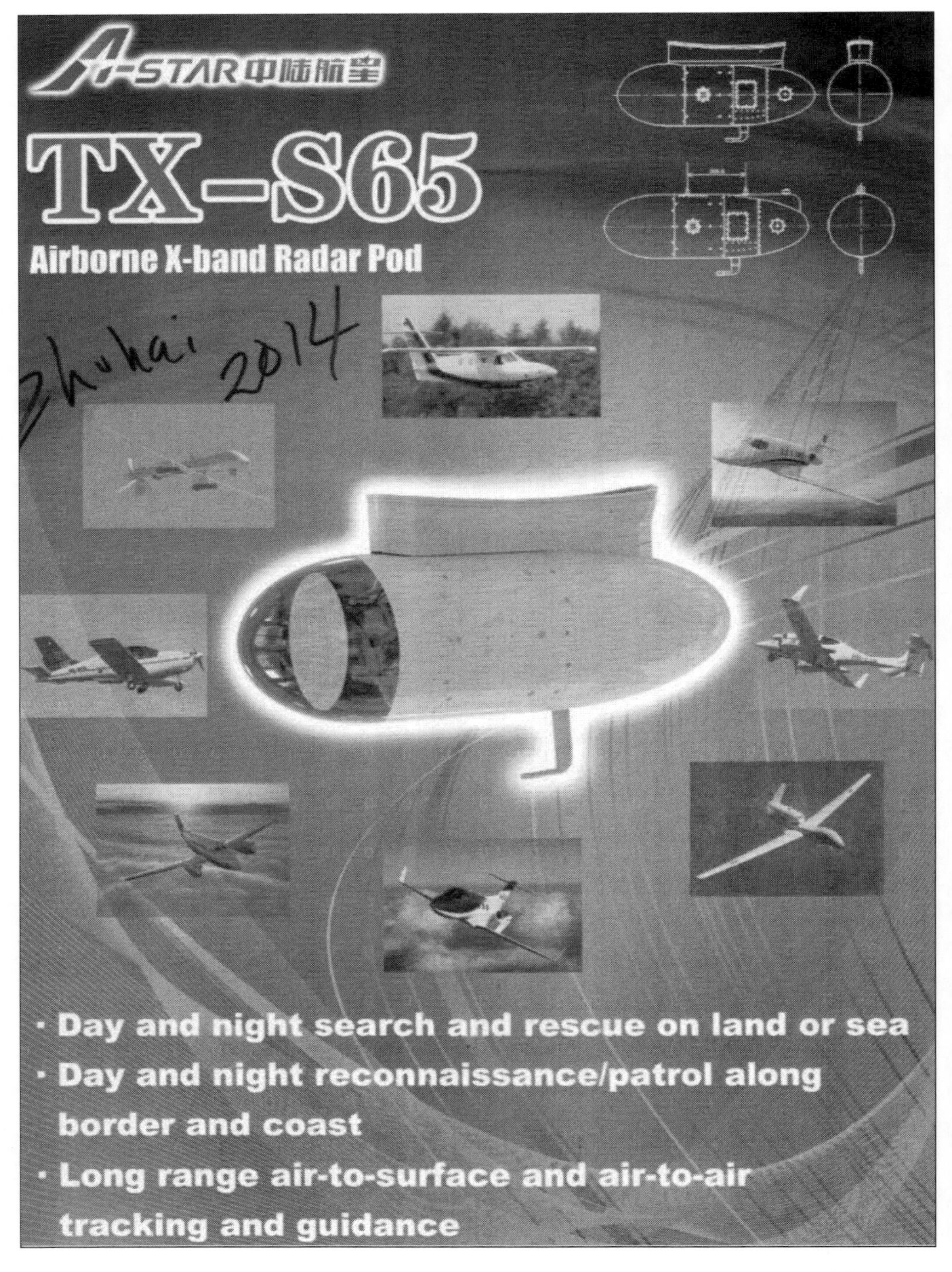

TX-S65. (A-Star) Airborne X-Band Radar Pod. 2014 Zhuhai Airshow (1/2).

TX-S65

A-STAR 中陆航星

Airborne X-band Radar Pod

Brief

TX-S65, a newly designed airborne radar pod with built-in X-band radar sensor, is suited for long-range missions of air-to-surface and air-to-air search, reconnaissance, tracking and guidance.

Applications

- Day and night search and rescue on land or sea
- Day and night reconnaissance/patrol along border and coast
- Long range air-to-surface and air-to-air tracking and guidance

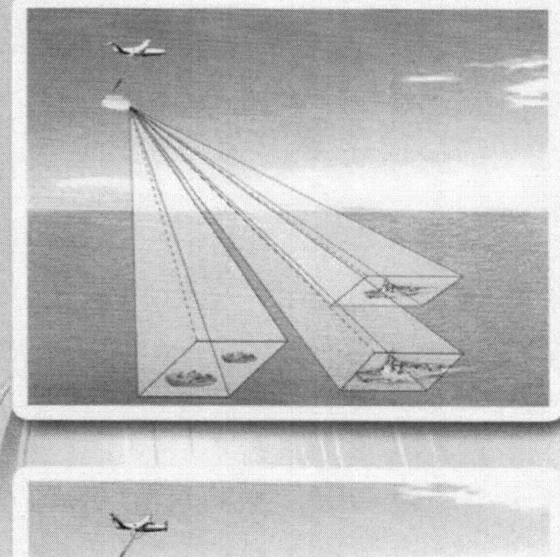

Performance benefits

- Carbon fiber composite pod, compact and light (55kg), small wind resistance, advanced performance
- Easy installation and well fitness to aircraft
- High performance-to-price ratio
- Long operating range, strong capacity of anti-sea clutter
- All solid-state units, easy modification, suited for a variety of aircraft models

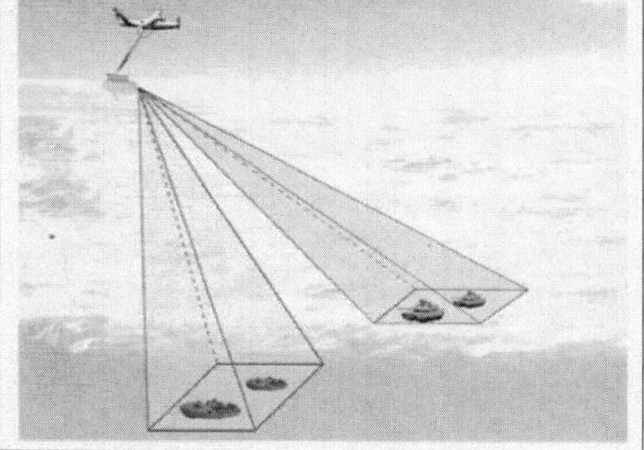

TX-S65. (A-Star) Airborne X-Band Radar Pod. 2014 Zhuhai Airshow (2/2).

TX-S320. (A-Star) Airborne Phased Array Radar Pod. 2014 Zhuhai Airshow (1/3).

TX-S320

A-STAR 中陆航星

Airborne Phased Array Radar Pod

Brief

Airborne phased array radar pod TX-S320 with build-in airborne phased array radar performs long-range air-to-surface and air-to-air search, reconnaissance and tracking tasks, it can also draw high-resolution SAR images of ground surface or islands.

Performance benefits

- Carbon fiber composite pod, compact and light (≤55kg), small wind resistance
- Phased array system with advanced performance
- Multi-function, multi-target and multi-purpose radar

Applications

- Day and night search and rescue on land or sea
- Day and night reconnaissance/patrol along border and coast
- Electronic interference

Function and operating mode

- Air-to-air search, indication and tracking
- Air-to-sea search, indication and tracking
- SAR target recognition method
- Front- and side-looking swath imaging function
- Ground moving target indicator (GMTI)
- Low resolution mapping image
- Anti-sea clutter, anti-active jamming
- Self-adaption (adaptive zero quantity ≧ 3)
- Channel calibration and self-test function

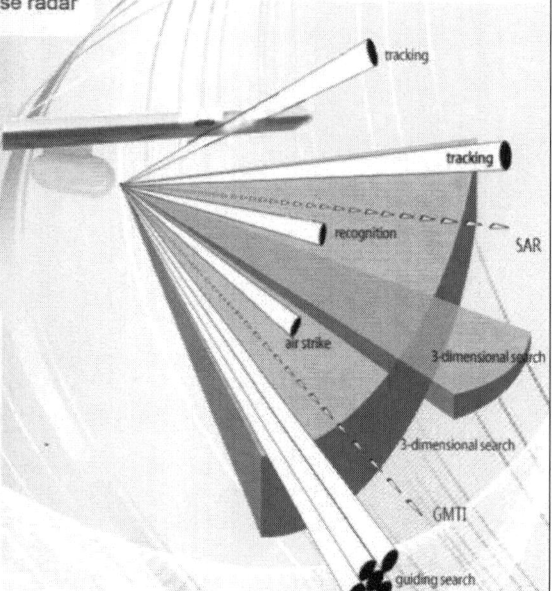

TX-S320. (A-Star) Airborne Phased Array Radar Pod. 2014 Zhuhai Airshow (2/3).

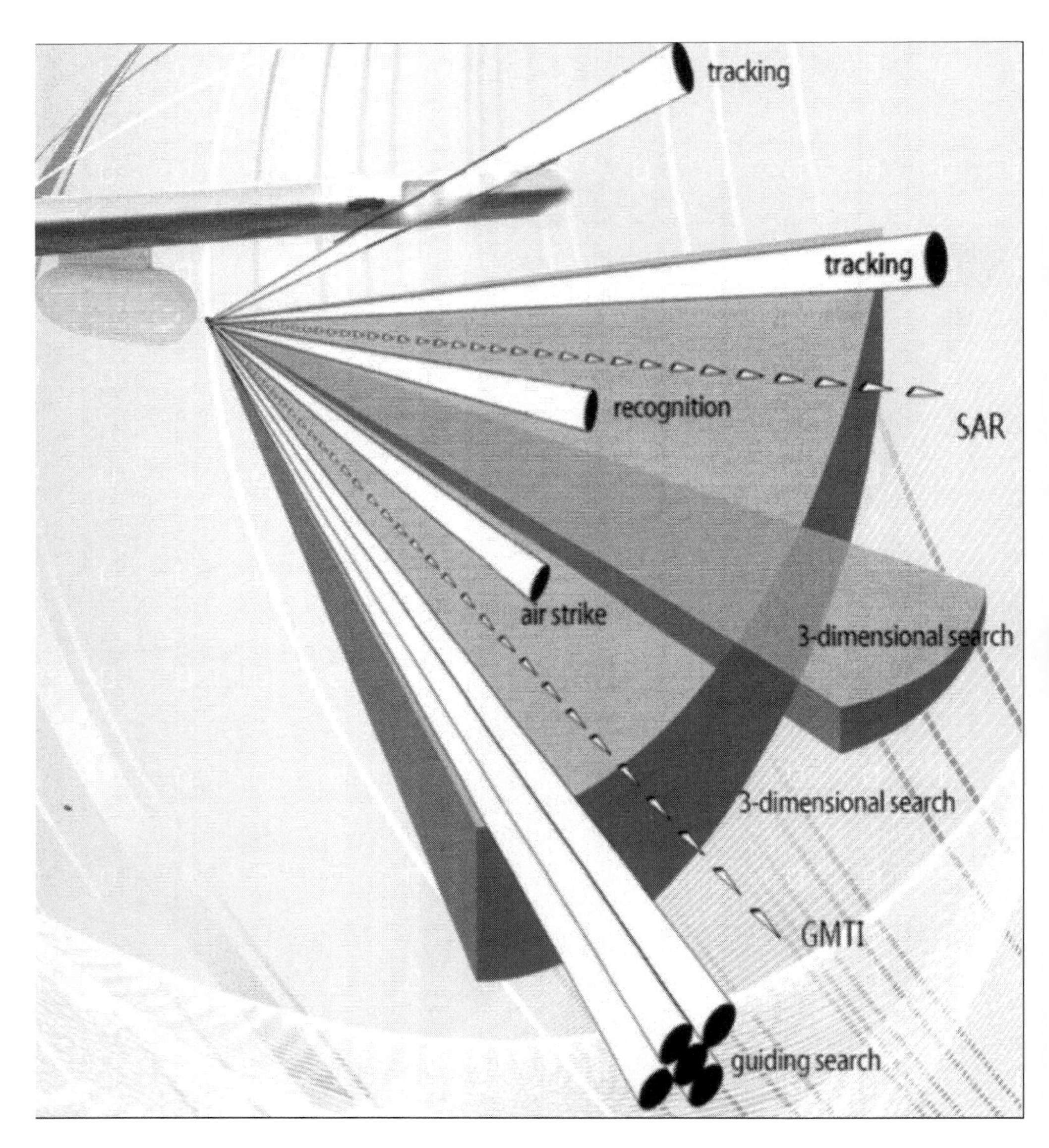

TX-S320. (A-Star) Airborne Phased Array Radar Pod. 2014 Zhuhai Airshow (3/3).

LASER/IR TARGETING POD

TYPE OC2

Zhuhai 2012

OC2激光红外照射吊舱

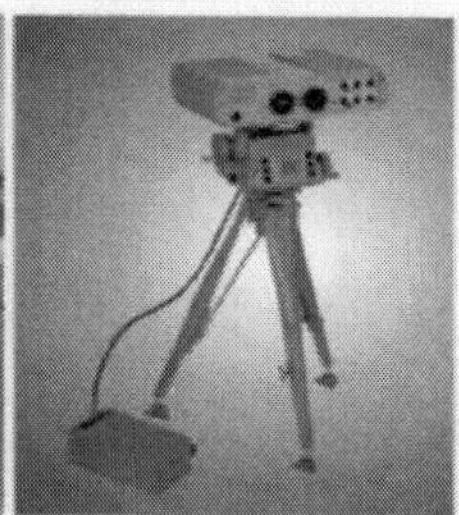

●USE AND CHARACTERISTICS

Laser Targeting Pod Type OC2 is a day/night targeting pod, equipped with an additional mid-wave FLIR (Front-Look-IR). FLIR is composed of head reflecting mirror, fixed reflecting mirror, three-in-one telescope system, light splitter 1, wide F.O.V. infrared imaging object lens, narrow F.O.V. infrared imaging object lens and infrared detector.

The pod has the functions of searching, detecting, identifying, tracking and illuminating ground targets, thereby, to guide the bomb to strike the target precisely at day/night and in bad weather conditions.

●MAIN TECHNICAL SPECIFICATION

Recognition range: 20 km

Max. illuminating range: 15 km

○Laser range finder:

Max. ranging: 20km

Range finding accuracy: <10 m

中国兵器工业集团 哈尔滨建成集团有限公司
NORINCO GROUP HARBIN JIANCHENG GROUP CO., LTD.

TYPE OC2 Laser/IR Targeting Pod. Norinco Group Harbin Jiancheng Group Co. 2012 Zhuhai Airshow (1/1).

"YINGS- Ⅰ" Infrared Searching and Tracking Pod

YINGS- Ⅰ Infrared Searching and Tracking Pod is capable of detecting, tracking, positioning targets in complex electromagnetic combating environment, which is not easy to be electronically jammed, and characterized by passive detection, good concealment, high positioning accuracy and strong capability of probing stealth targets. It could improve the ability of carrier aircraft to detect hidden targets in complicated electronic warfare environments, night and adverse weather condition.

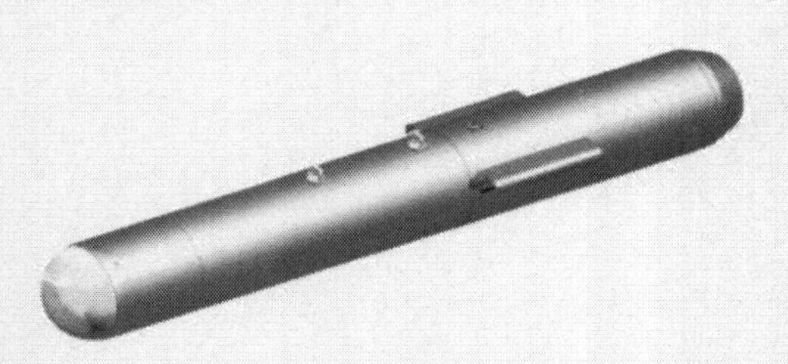

Functions

- Autonomously perform target searching in designated airspace
- Implement target detection in designated airspace according to external guidance information
- Capable of tracking one or two targets in the air
- Able to output targets angle information

Performance Specification

- IR Operating Band: Long Wave 7.7μm~10.3μm
- Number of Tracking Targets: In TWS mode, be capable of tracking two batches of targets at the same time
- Max searching/tracking range :
  Azimuth -60°~+60°
  Elevation -15°~+60°
- Size and Weight: Maximum size Φ300 ×2400mm
  Maximum weight: 130kg
- External electrical interface: MIL-STD-1553B; MIL-STD-1760C

AVIATION INDUSTRY CORPORATION OF CHINA Luoyang Institute of Electro-optical Equipment
Tel:0379-63327181 Email: eoei@vip.sina.com

Yings-I Infrared Searching and Tracking Pod. Aviation Industry Corporation of China (AVIC). Luoyang Institute of Electro-Optical Equipment. Unknown Defense Show (1/1).

"YINGS–II" Forward Looking IR

"YINGS-II" Forward Looking IR (FLIR) is a high-resolution scanning IR imaging system stored on fighter; it is mainly used for dual-FOV detection & imaging of targets in front of aircraft at night or in adverse weather condition, and outputting IR video displayed on Head-up Display (HUD) or Head-down Display (HDD), for the pilot to timely realize the terrain condition of flight course ahead, as well as assisting in taking-off/landing and low-altitude penetration, and guiding targeting pod to quickly find the target.

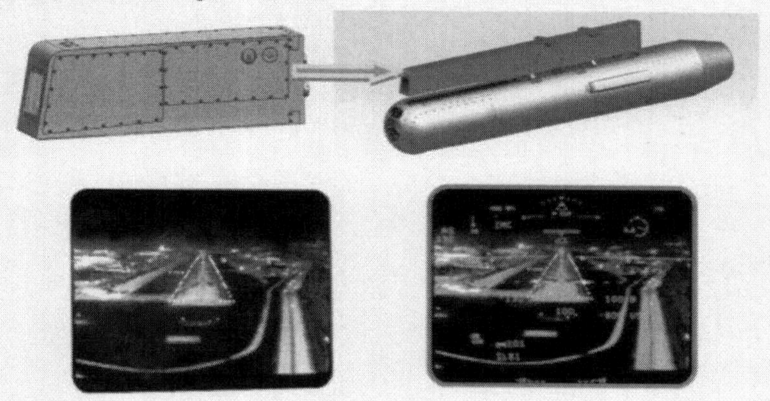

Functions

- To assist in taking-off/landing at night or in adverse weather condition
- To assist aircraft with low-altitude penetration at night by overlapping video on HUD or HDD
- To provide IR WFOV to targeting pod, and guide pod to find target

Specification

- WFOV: 18.3°×13.7°~22.7°×17°(azimuth × elevation)
- NFOV: 16°×12°(azimuth × elevation)
- Max. observation range: 51°×25.5°~64°×32°(T shape)
- Operation waveband: 3μm~5μm
- Max outline dimension: 600mm ×110mm×170mm (Length × Width × Height)
- Max weight: 15kg

Yings-II Forward Looking Infrared. Aviation Industry Corporation of China (AVIC). Luoyang Institute of Electro-Optical Equipment. Unknown Defense Show (1/1).

YINGS–III330 Day/Night Targeting Pod

With common aperture optical design technique, YINGS-III 330 day/night targeting pod is able to significantly minimize dimension and weight of targeting pod; stealth and aerodynamic performance could be upgraded according to adoption of cuneal joint electro-optical window configuration.

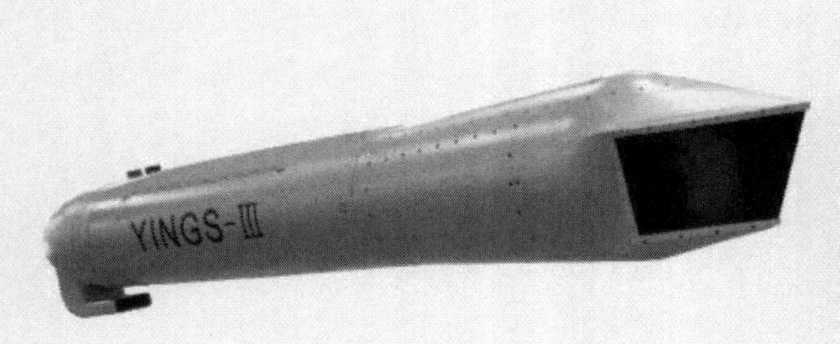

Functions

- To perform searching, detection, identification and tracking on ground/maritime target with IRST and TV sensor in day/night;
- To acquire distance and angle information of target with laser ranging function; and direct laser-guided weapon to attack the target with laser designation;
- To provide image support for target damage effect evaluation.

Specifications

- Moving scope: Azimuth: ±150°; pitch: +10°~−150°
- Outline dimension: Head diameter≤Φ360mm, diameter of other segments≤Φ330mm length≤2300mm
- Weight: ≤230kg
- IR / TV WFOV: 3.6°×2.7°; IR / TV NFOV: 1.2°×0.9°
- External electrical interface:
  PS: DC 28V; three-phase:115V/400Hz
  Bus: 1553B bus
  Video: PAL composite video /STANAG SS3350B video

Yings-III 330 Day/Night Targeting Pod. Aviation Industry Corporation of China (AVIC). Luoyang Institute of Electro-Optical Equipment. Unknown Defense Show (1/1).

YINGS–III 360 Day/Night Targeting Pod

With optimization of system distribution, YINGS-III 360 day/night targeting pod adopts optical-mechanical design technique with high volumetric ratio to decrease dimension and weight of product significantly, so as to enable it to be applicable for light aircraftsas trainer, middle and small size of fighter.

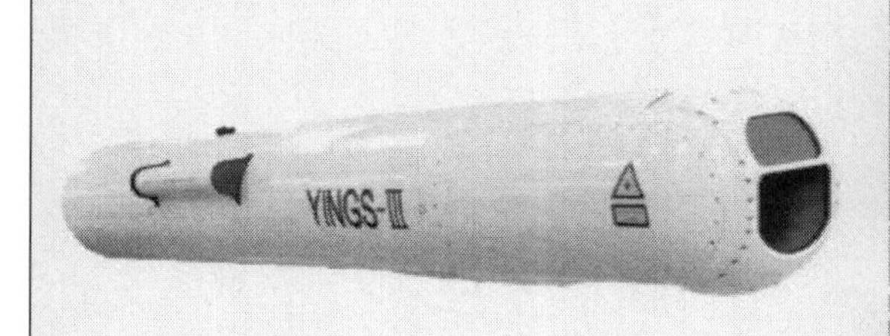

Functions

- To perform searching, detection, identification and tracking on ground/maritime target with IRST and TV sensor in day/night
- To acquire distance and angle information of target with laser ranging function; and direct laser-guided weapon to attack the target with laser designation
- To provide image support for target damage effect evaluation

Specification

- Movement scope: Azimuth: ±150°; elevation: +10°～－150°
- Outline dimension:
  Head diameter≤Φ360mm, diameter of other segments≤Φ330mm,length≤2400mm
- Weight: ≤240kg
- IR / TV WFOV: 3.6°×2.7°; IR / TV NFOV: 1.2°×0.9°
- External electrical interface:
  PS: DC 28V; three-phase:115V/400Hz
  Bus: 1553B bus
  Video: PAL composite video /STANAG SS3350B video / digital video

Yings-III 360 Day/Night Targeting Pod. Aviation Industry Corporation of China (AVIC). Luoyang Institute of Electro-Optical Equipment. Unknown Defense Show (1/1).

YINGS–III 390 Day/Night Targeting Pod

As an airborne electro-optical detection system integrated with IR, TV and laser, YINGS-III 390 day/night targeting pod is able to perform searching, identification and tracking on target in day/night; as well as to guide laser-guided bomb to implement attacking precisely.

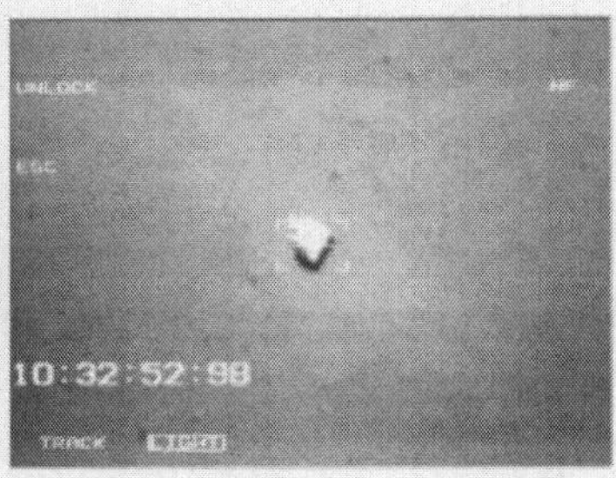

Functions

- To perform searching, detection, identification and tracking on ground/maritime target with IRST and TV sensor in day/night
- To acquire distance and angle information of target with laser ranging function; and direct laser-guided weapon to attack the target with laser designation
- To provide image support for target damage effect evaluation

Specifications

- Movement scope: Azimuth: ±150°; elevation: +10°~−150°
- Outline dimension: Φ390mm×2700mm
- Weight : 280kg
- TV WFOV: 4.0°×3.0°; TV NFOV: 1.0°×0.75°
- IR WFOV: 4.0°×3.2°; IR NFOV: 1.0°×0.8°
- External electrical interface:
  PS: DC 28V; three-phase 115V/400Hz
  Bus: 1553B bus
  Video: PAL composite video or STANAG SS3350B video

Yings-III 390 Day/Night Targeting Pod. Aviation Industry Corporation of China (AVIC). Luoyang Institute of Electro-Optical Equipment. Unknown Defense Show (1/1).

WMD-7 Day/Night Targeting Pod

WMD-7 model day/night targeting pod is an airborne electro-optical detection system, with IR, TV and laser sensors integrated inside. With IR and TV sensors, ground targets can be searched, recognized and identified in day/night conditions. In the tracking status, laser can be used to range and designate targets, and guide precise guided weapons, such as laser guided bombs, or common bombs to hit and bomb enemy targets precisely.

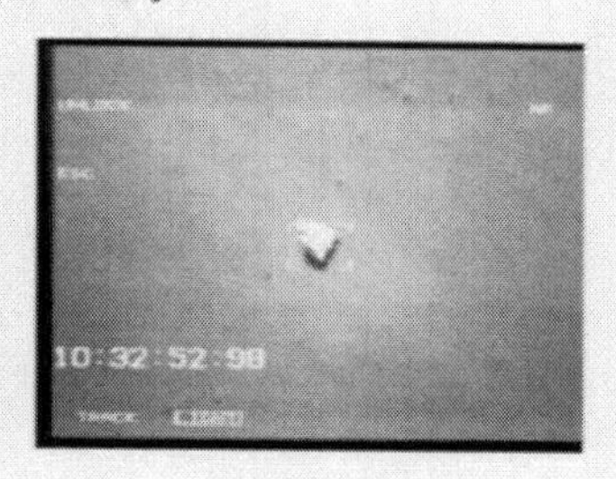

Functions

- Ground or sea targets can be searched,recognized and identified in dag and night conditions
- Automatic tracking targets after identified
- With laser range-finder, providing range and angle information of targets for fire Control system
- With laser designation, guiding laser uided weapons to hit targets precisely
- Image support for damage Assessment

Sensor Configuration

- Movement range: Azimuth:±150°; Elevation: +10°～−150°
- Dimension: Φ390mm×2700mm
- Weight: 280kg
- TV
  Wavelength: 0.6～1.0μm
  FOV: WFOV: 4.3°×5.8°; NFOV: 1.4°×1.9°
- IR
  Wavelength: 3～5μm
  FOV: WFOV: 4.3°×5.8° ; NFOV: 1.4°×1.9°
- Laser
  Wavelength: 1.064μm
  Maximum laser designating range: ≥13km
  Laser ranging distance: 0.5km～18km
- The external electrical interface
  Power Supply: DC 28V; Three-phase 115V/400Hz
  Bus: 1553B bus
  Video signal: Composite video signal of PAL system

AVIATION INDUSTRY CORPORATION OF CHINA Luoyang Institute of Electro-optical Equipment
Tel:0379-63327181 Email: eoei@vip.sina.com

WMD-7 Day/Night Targeting Pod. Aviation Industry Corporation of China (AVIC). Luoyang Institute of Electro-Optical Equipment. Unknown Defense Show (1/1).

WMD-7 Day/Night Targeting Pod

WMD-7 model day/night targeting pod is an airborne electro-optical detection system, with IR, TV and laser sensors integrated inside. With IR and TV sensors, ground targets can be searched, recognized and identified in day/night conditions. In the tracking status, laser can be used to range and designate targets, and guide precise guided weapons, such as laser guided bombs, or common bombs to hit and bomb enemy targets precisely.

Composition

- Head segment
- Rolling segment
- Environment control Segment
- Electronic bay segment

Specifications

- Movement range:
  Azimuth: ±150° ;
  Elevation: +10° ~ −150° .
- Dimension: Φ390mm×2700mm.
- Weight: 280kg.
- TV
  Wavelength: 0.6~1.0 μm;
  FOV:
  WFOV: 4.3° ×5.8° ;
  NFOV: 1.4° ×1.9° .
- IR
  Wavelength: 3~5 μm;
  FOV:
  WFOV: 4.3° ×5.8° ;
  NFOV: 1.4° ×1.9° .
- Laser
  Wavelength: 1.064 μm;
  Maximum laser designating range: ≥13km;
  Laser ranging distance: 0.5km~18km.
- The external electrical interface :
  Power Supply: DC 28V; Three-phase 115V/400Hz;
  Bus: 1553B bus;
  Video signal: Composite video signal of PAL system.

Functions

- Ground or sea targets can be searched, recognized and identified in dag and night conditions;
- Automatic tracking targets after identified;
- With laser range-finder, providing range and angle information of targets for fire Control system;
- With laser designation, guiding laser guided weapons to hit targets precisely;
- Image support for damage Assessment.

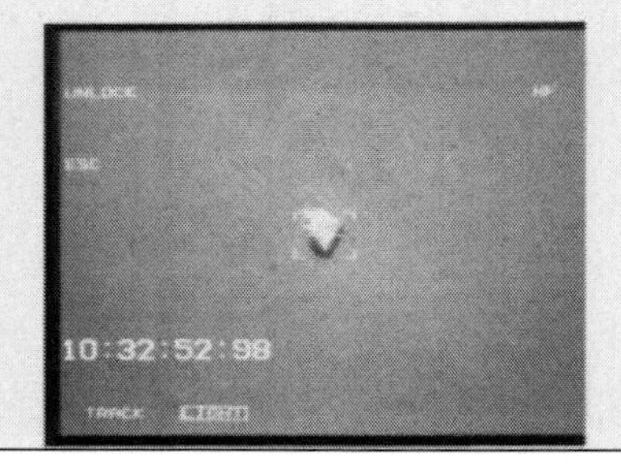

WMD-7 Day/Night Targeting Pod. Aviation Industry Corporation of China (AVIC). Luoyang Institute of Electro-Optical Equipment. 2012 Zhuhai Airshow (1/1).

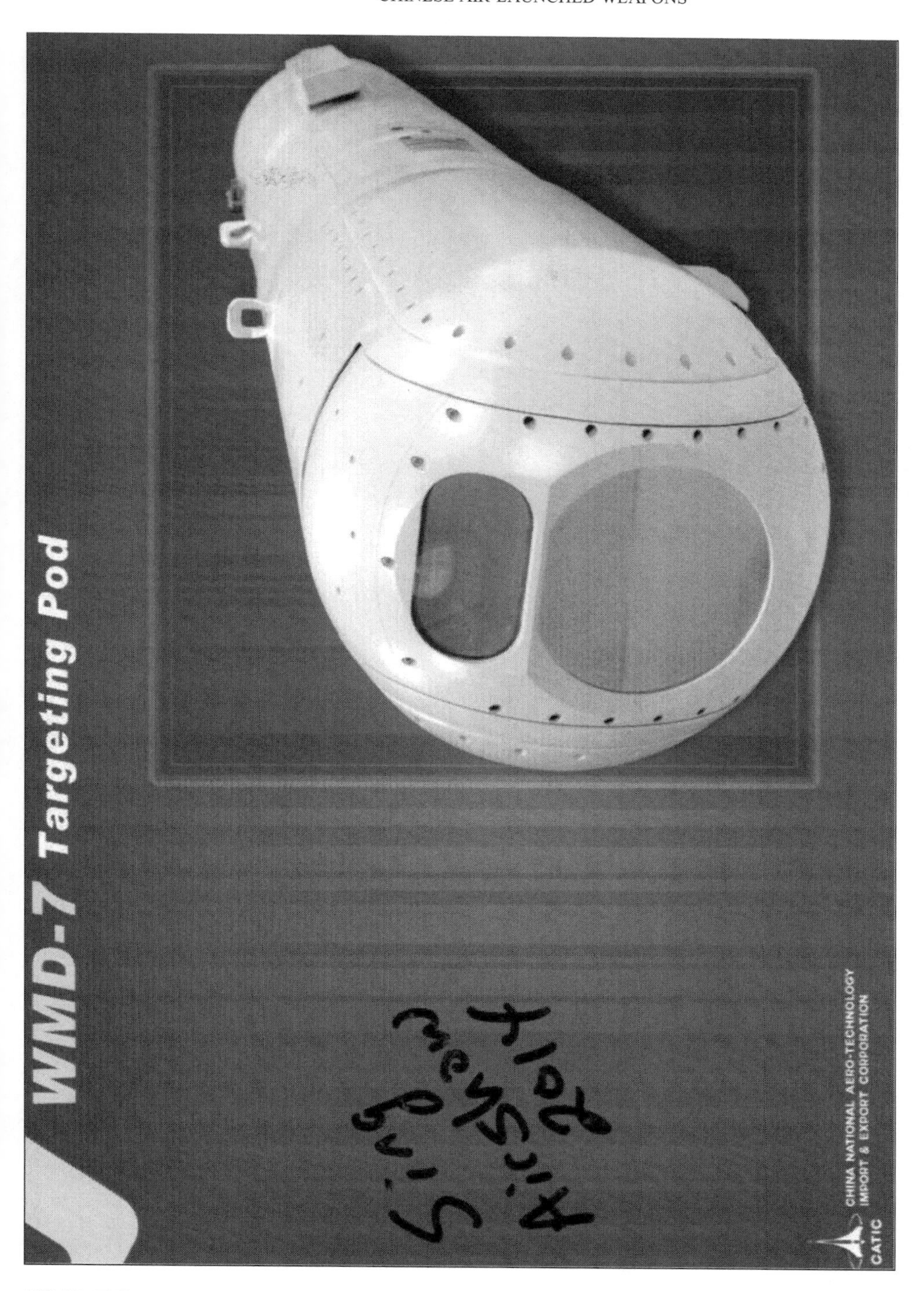

WMD-7 Targeting Pod. China National Aero-Technology Import and Export Corporation (CATIC). 2014 Singapore Airshow (1/3).

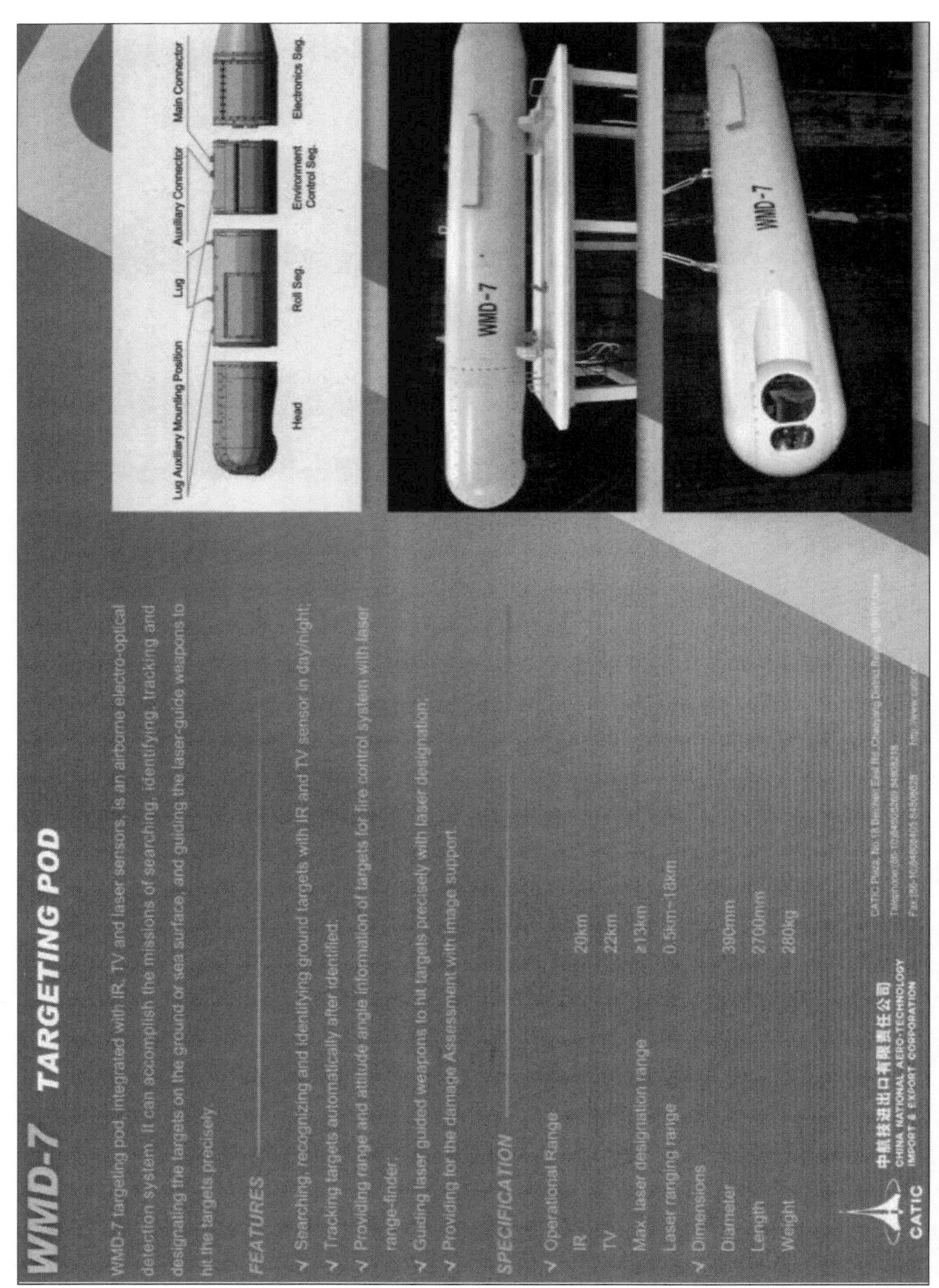

WMD-7 Targeting Pod. China National Aero-Technology Import and Export Corporation (CATIC). 2014 Singapore Airshow (2/3).

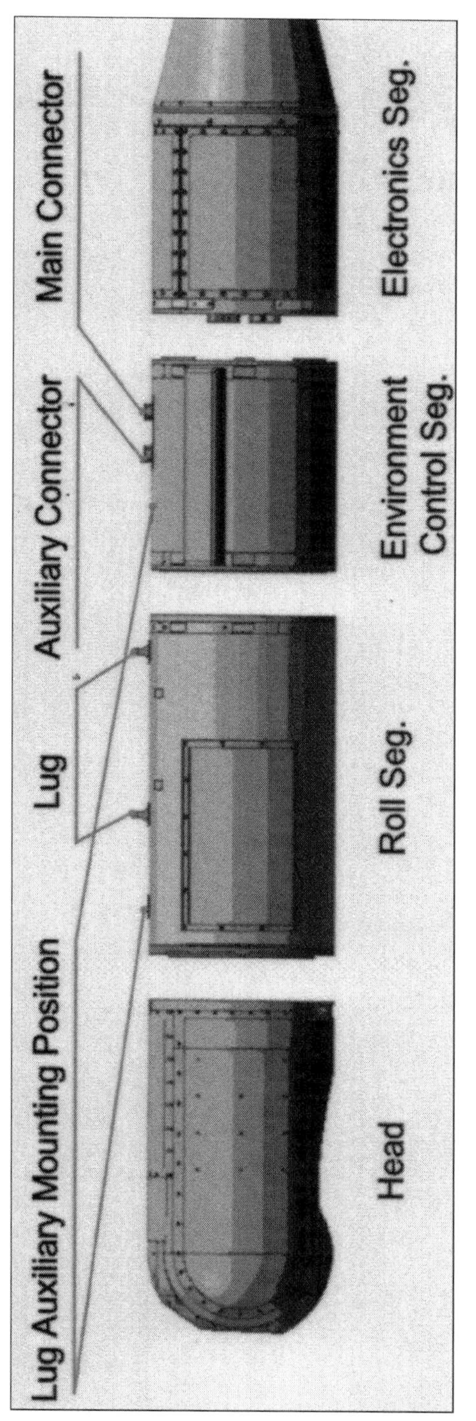

WMD-7 Targeting Pod. China National Aero-Technology Import and Export Corporation (CATIC). 2014 Singapore Airshow (3/3).

INDEX

CHINA MARKET OUTLOOK FOR CIVIL AIRCRAFT, 2014-2033

INCLUDING COMMERICAL AIRCRAFT PRODUCT BROCHURES

WENDELL MINNICK
EDITOR

Taiwan Space Vehicles: Documents
ORBITAL
NSPO
ASTRIUM
XL FIRST FLIGHT
Wendell Minnick

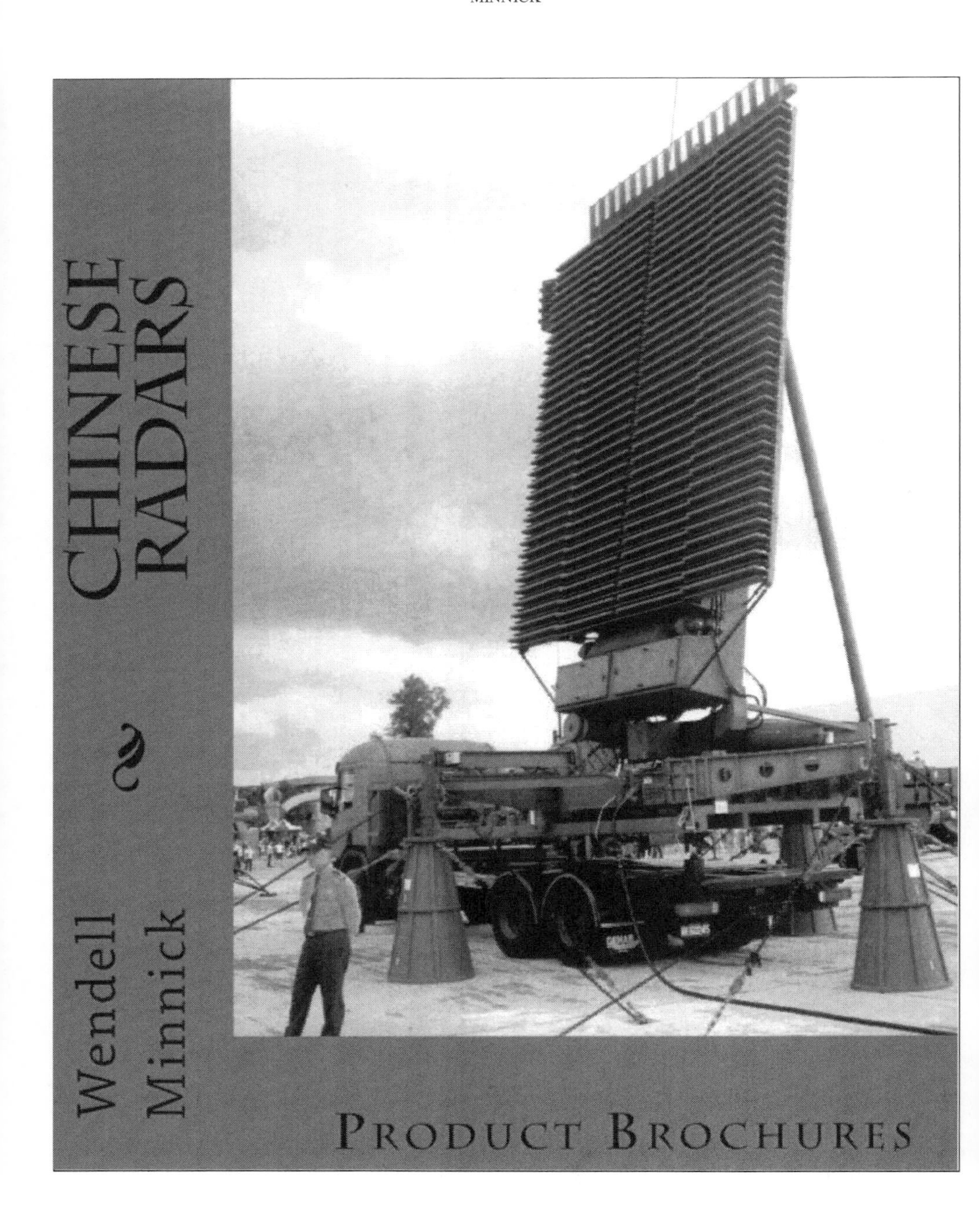
CHINESE RADARS
Wendell Minnick
PRODUCT BROCHURES

CHINESE SEAPLANES, AMPHIBIOUS AIRCRAFT AND AEROSTATS/AIRSHIPS

- PRODUCT BROCHURES -

WENDELL MINNICK
EDITOR

CHINESE FIXED-WING UNMANNED AERIAL VEHICLES

RODUCT BROCHURES

WENDELL MINNICK

EDITOR

CHINESE ROCKET
SYSTEMS
MULTIPLE LAUNCH ROCKET SYSTEMS
BROCHURES
EDITED BY
WENDELL MINNICK

CHINESE SPACE VEHICLES AND PROGRAMS

Wendell Minnick
Editor

- PRODUCT BROCHURES -

TAIWAN CYBERWARFARE
國防部參謀本部資通電軍指揮部網路戰聯隊
INFORMATION COMMUNICATIONS AND ELEC ONIC FORCE COMMAND CYB WARFARE WING
WENDELL MINNICK

Chinese Tanks and Mobile Artillery
BROCHURES
WENDELL MINNICK
EDITOR

DIRECTORY OF FOREIGN AVIATION COMPANIES IN CHINA

COMMERCIAL AND DEFENSE

WENDELL MINNICK

Printed in Great Britain
by Amazon